PRIME CONTRACT

Building the Navy's SEANET System

A Novel

By

David L. Lyndon

CHAPTER ONE
PROTEST

Power in the United States Congress is concentrated in the permanent committees, whose function is to investigate and contemplate relevant issues and to frame legislation for the parent body's consideration and action. While one might quarrel with the efficiency and fairness of this process, it has, all things considered, worked rather well over two centuries, although notable exceptions exist.

Of the committee chairs, the Chairman of the Senate Armed Services Committee, SASC, wields extraordinary power. Together with his counterpart in The House of Representatives, the Chairman of the House Armed Services Committee, HASC, their committees have control over the armed forces' funding and how those funds are spent. With the President empowered as Commander in Chief, civilians, the people's elected representatives, are clearly in charge of the U.S. military -- precisely as intended by the framers of the Constitution.

The framers did not anticipate, however, the evolution of the committee system and the Honorable Joseph T. Flannery, senior senator from the Commonwealth of Massachusetts. Having been named, in the middle of his fourth six-year term, Chairman of the Senate Armed Services Committee, he was the disproportionately influential member of the civilian troika charged with the nation's defense. Little could happen in the military without his approval, and that had to be obtained at a political cost almost without exception. Bartering favorable decisions for favors in kind, while threatening retribution on the constituency of uncooperative senators, he exercised virtual veto power over matters of consequence to the armed services. The closing of a military base, the award of a contract, the funding of a project, the approval of an appointee to a senior Department of Defense post or the promotion of an officer to the rank of Admiral

1

or General, these were objects of political barter traded by Flannery, and he made no bones about it. Flannery did not win every battle, but he did win most of them, and he always stayed on the right side of the law while relying on the ambiguity of congressional ethics. Nevertheless, occasionally the odds were just too great or his cause too unpopular to carry the day. "Politics is politics" was not only the Flannery motto, but also the requested epitaph to be inscribed at his final resting place -- assuming, of course, that he could not cut a deal for immortality. His countless enemies opined that he was working on it, but hoped, nevertheless, that the intelligent citizens of Massachusetts would find disfavor with J.T.F. when next they were asked to return him to Washington for six more years even if a bargain had been struck with the dark angel.

iiiiiiiiiiiiiiiiiiiiiiiiiiiiiiiiiiiiiiiiiiiiiiiiiiiiiiiiiiiiiiiiiiiiiiiiiiiiiiiiiiiiiiiiiiiiiiiiiiiiiiiiiiiiiiiiiiiiiiiiiiiiiiiiiiiiiiiiiiiiiiiiiiiiiiiiiiiiiiiiiiiiiiiiiiiii

The competition for developing SEANET, an acronym for Space Environment Automated Naval Electronics Transmission, began in 1993 and had been hard fought. Major military contractors across the country had competed for it over a period of five years. First there were the engineering study contracts in which various concepts were put forward by the industrial combatants, followed by a down selection to three contractor teams for development of detailed designs and engineering model hardware, and then the final selection -- winner take all. The three finalists were Tellonics in California, Electrospace Corporation of Texas, and New England Electronics in Massachusetts. The latter, Massachusetts' foremost aerospace company, had submitted the lowest cost estimate for the project by some five percent. However, the Navy's Source Selection Board found its technical approach second best and its cost estimates questionable. After months of evaluating as objectively as possible the three proposals and the results of the technical "flyoff" of the three contractors' engineering models, the board made its recommendation to the Source Selection Authority, Rear Admiral Russell "Rusty" Sullivan. Sullivan concurred with the board's recommendation, and Tellonics was selected as the prime contractor for SEANET. On a Friday afternoon in March of 1998, as is customary, the members of the House and Senate from the State of California were notified so that they could make the initial public announcements and issue press releases in order to accumulate political capital at home for a California firm having been selected. The Navy would announce the winner informally through Admiral Sullivan's personal phone calls to the CEOs of each of the competing contractors. The competitors would receive formal written notification in a few days.

Flannery was annoyed that he had not been called by mid-afternoon, and upon learning that New England Electronics had not been selected, he was beyond furious. Massachusetts' Representatives and Senators had lobbied hard for their state's contractor. His immediate response was to contact Ralph Lowry, CEO of New England Electronics, and demand that the award be formally protested. Lowry agreed, and would, in fact, have done so without prompting from Flannery. He instructed his staff to contact the Navy and request a formal debriefing on the source selection, a necessary first step of the protest process. Once the formal debriefing was conducted, the competitors had ten days to file a formal protest with GAO, the General Accountability Office of the United States Congress.

Rear Admiral Sullivan had anticipated that a debriefing would be requested and had already made arrangements for it to take place on the following Wednesday. The protocol of such briefings was carefully constructed to avoid confrontation between the evaluators and the evaluated. Presentations would be made by various Navy personnel to disclose the results of the competition and the rationale for having selected Tellonics. There would be no questions allowed and no discussion. The competitors were invited to submit written questions in advance, although the Navy was under no obligation to answer them.

In major competitions such as SEANET, the Request for Proposals, the RFP, is voluminous, in this case over 500 pages of specific requirements including enumeration of a multitude of Federal Acquisition Regulations, FARs, incorporated by reference. The FAR volumes occupy a full six feet of shelf space. The technical specifications were a separate secret document of another 300 pages, and there was a twenty-five page top-secret-crypto addendum specifying encryption requirements. A critical component of the RFP was right up front – the evaluation criteria. These criteria were spelled out in detail to make clear how the proposals would be evaluated. Each criterion was assigned a maximum weight, or percentage, that could be awarded for that element. For SEANET up to 40 points could be assigned for technical approach and company qualifications, 30 points for cost effectiveness, and the remaining 20 points for various support elements – reliability, maintainability, availability, producability, supportability – the "ilities" as they are called in the aerospace industry. The competitor with the highest overall score becomes the winner. One difficulty, of course, is that the ratings are assigned subjectively by the specialists assigned to the Source Selection Board.

Representatives of the competing contractors assembled at the designated Space and Naval Warfare Systems Command, SPAWAR, conference room at 1300 hours on Wednesday. The debriefing disclosed that Tellonics overall score was 85 points, New England Electronics 80 points, and Electrospace 70 points. The heaviest weight was assigned to technical approach, and in that category Tellonics scored 35 points versus New England Electronics 30 points. On cost, both received 25 points, and the other factors were also a wash in total with 15 points each. A Source Selection Board specialist for each of the categories presented the rationale for the points assigned to each competitor.

New England Electronics suspected that something was wrong. The dollar amount of the award to Tellonics, $1,085,977,327, was now a matter of public record. New England Electronics' proposal, still proprietary and not made public, had been almost 50 million dollars lower, nearly five percent less at $1,037,108,343. How could it be that both were scored the same 25 points on cost effectiveness? The debriefing specialist for cost effectiveness had stated only that the cost proposals were adjusted by the evaluators to take into account the relative risk of the technical proposals and the cost estimating methodology employed by the bidders. The cause of a protest was at hand.

On a Friday afternoon nine days later the General Accountability Office logged in a hand carried letter from New England Electronics formally protesting the award, accompanied by a legal brief containing the rationale for the protest, and requesting an immediate hearing. Protest to the GAO was the first of several appeals that a contractor could make in the event an acquisition decision did not favor it, and protest based on cost could very well succeed. Since the protest had been filed with GAO within ten days of the award debriefing, the contract to Tellonics could not be awarded without GAO approval.

Although an agency of The Congress empowered to act independently, GAO was not immune to political pressure from the members, and Senator Flannery knew it. The Chairman of the House Armed Services Committee, Elias James, knew it too. James, no political novice, was the Representative of the 12th congressional district of California.

The Comptroller General of the United States, Keith Majors, the senior official of the GAO, received two important telephone calls on the day the protest was received by GAO. The first was from HASC Chairman James. Majors answered,

5

"Good afternoon, Mister Chairman. How can I be of service so late on a Friday afternoon? Are you about to spoil another weekend?"

"Not at all, Keith, at least I hope not." James replied. "Just wanted to bring to your attention the fact that a billion dollar contract is about to be awarded to Tellonics for the SEANET project. Naturally, we Californians are well pleased with that decision, but I suspect that our colleagues from Massachusetts might feel otherwise. You may be hearing from one or more of them, because I understand a protest has been filed."

"Well, congratulations, Mr. Chairman, to you and your fellow Californians. And yes, that is the case. A protest has been filed. We'll do our best to insure a fair and just outcome as we always try to do."

James, anticipating that Flannery might attempt to pressure Majors into assigning a chief investigator partial to New England Electronics, continued. "Of course you will. But you may be under a great deal of pressure, and I just wanted you to know how important it is in this case to assign your very best, objective people to the job. Let the chips fall where they may after that."

"That's very good advice, Mr. Chairman, and I'll certainly take it. Is there anything else I can do for you today?"

"That's about it, Keith. Enjoy your weekend," James concluded.

The phone call from Flannery followed soon after. "Good afternoon, Keith. This is Joe Flannery." They had known each other for over ten years. Each respected the other, but there had been some problems. Flannery tended to meddle in GAO affairs a bit too much for Majors' liking. The independence of the GAO, even though an agency of the Congress, was a sacred trust so far as Majors was concerned, not to be manipulated by individual members of congress."Hi, Joe. What's up?" Majors adopted a friendly and casual tone.

"Well, Keith, I'm sure you're aware that the SEANET project may be awarded to Tellonics out there in California. I got a call from Ralph Lowry at New England Electronics just now, and he informed me that he has filed a protest. I think he's got a lot of justification for getting this award overturned, so I thought I'd give you a heads up that trouble is coming your way."

"Trouble, Joe? How's that? We handle a lot of protests, you know, and we usually resolve them without too much difficulty." Majors wanted to appear unconcerned.

"There's been entirely too much Navy business going to California lately, and I'm beginning to suspect that there's an uneven playing field. I'd like for you to be pretty thorough on this investigation and find out what's really going on. You'd better dig pretty deeply into this one, and I suggest you assign a lead investigator from a New England state to be sure there's no California hanky-panky. I'll be watching this very closely, Keith. It's a bigger issue than just this one contract, so I might have to conduct hearings if things don't work out right." Flannery's warning was clear enough.

"Well, thanks for the heads up, Joe. I'll certainly take your concerns into account, and you needn't worry about thoroughness. I'll put a good team on this."

"That's fine, Keith, but let's stay in touch until this matter is satisfactorily resolved. I'm off to Cape Cod in a few minutes, but I'll call you next week sometime to see how things are progressing."

"Any time, Joe. Always a pleasure to talk with you," Majors concluded the exchange.

Flannery did not realize that Majors, contemplating an early retirement, was not susceptible to his threat to conduct congressional hearings if the award was not overturned. In fact what Flannery had inadvertently done was to arouse Majors' competitive spirit, which was considerable. As he prepared to depart his office for a well-deserved weekend of relaxation, he reached under the center desk drawer and switched off the telephone tape recorder.

+++++++++++++++++++

Marcia Surrett called the meeting to order at nine o'clock on the eighteenth of June in a first floor conference room in the Gerald Ford Office Building. The choice of Surrett to lead the protest inquiry was made in the light of her experience and reputation for being fair and unbiased. What's more, she was a native of North Carolina, not partial to either state of the adversaries. A senior civil servant of grade GS-15, a rank equivalent to Army Colonel or Navy Captain, she had a twenty-year career at GAO, starting with a law degree from Duke University and earning her MBA from Harvard University along the way. Present were her GAO legal and technical investigative team members. Both New England Electronics and Tellonics were represented by their lawyers, contract administrators, and proposal managers. The Navy was not represented. GAO had consulted them and would consult them again independently, thus shielding them from direct discourse with the contractors. A court stenographer was present to record the meeting with both a steno machine and audio tape recorder.

"Good morning, ladies and gentlemen, " Surrett began. "We are here today for verbal presentation of a contract award protest which was timely filed by New England Electronics regarding the United States Navy's SEANET competition. A legal brief has been received from New England Electronics detailing their cause of action. A copy of that brief was provided to Tellonics with proprietary and competition sensitive information redacted from that copy, and Tellonics has filed a timely response. A copy of that response, similarly redacted, was provided to New England Electronics. All of these documents have been reviewed by the appropriate GAO

8

personnel, and I have studied them personally. We have also met with the Source Selection Board and the Source Selection Authority, but only for the purpose of fact-finding and collection of pertinent solicitation documents. Today's meeting will conclude the information-gathering phase of the protest process. GAO will then retire to consider the merits of the protest, and although we are allowed a great deal more time to do so, we expect to announce our decision within a week or two since the Navy is anxious to proceed with this program."

"If there are no objections, I call on New England Electronics for any comments or presentation they may wish to make."

Richard Owens, legal counsel for New England Electronics, stood. "Good morning, Ms. Surrett." Owens was careful with his pronunciation of Mizz Surrett. "Our brief set out the arguments that we contend merit a contract award to New England Electronics rather than Tellonics. I would like to go through our argument with you in very abbreviated form, stating its principal points, in order that GAO have an opportunity to inquire about any of those points if they are not understood or require elaboration. We shall be delighted to clarify any of those points with you.

"I note, Ms. Surrett, that members of the competition are present. I would like to request that they be excluded from the meeting during our presentation since it will be necessary for us to present competition sensitive information."

"Fair enough, Mister Owens. However, if you raise any additional points of argument during your presentation that were not covered in your legal brief, then I am obliged to bring those supplemental arguments to the attention of Tellonics' representatives here today so that they may respond in kind. Is that satisfactory?" Surrett's reputation for fairness was demonstrated clearly. There would be no tricks played here. Then, "Is that agreeable to you Mister McBride?" this directed to

9

the Vice President of Contract Administration for Tellonics, Albert McBride.

"Yes, that is fine with us. I would also appreciate an opportunity to make a few remarks after we return if that is permissible," replied McBride.

"By all means, Mr. McBride. So if you would be so kind, will you and your colleagues please excuse us for a while? You'll find a comfortable break room just down the hall to your left, and we'll call you back in shortly." Surrett was not only fair, but also sensitive to the dignity of the parties involved in the dispute.

Owens proceeded to summarize the New England Electronics appeal, point by point. He was an eloquent speaker, obviously an experienced trial lawyer, and his argument was compelling. Its main point: why was the award not made to the low bidder when both competitors' approaches were technically compliant with the requirements of the solicitation? Why was the Navy wasting millions of dollars of taxpayer money? Although attentive to every word, there were few questions from GAO, and no new points raised by Owens.

The presentation lasted about forty minutes, after which Surrett suggested they take a short break and reconvene at ten fifteen.

Surrett resumed the meeting after the break with all parties present. "New England Electronics has made its verbal presentation, and GAO is satisfied that it understands the nature of the award protest quite clearly. No new points were raised beyond those contained in the New England Electronics brief; so at this time, Mr. McBride, assuming you would like to offer comments, would you like to do so without New England Electronics present?"

"Yes, we would like to comment, Ms. Surrett, and thank you very much for the opportunity. It's not necessary for New England Electronics to leave. In fact, I sincerely hope that our counter-argument will be so convincing that they will reconsider their protest." The New England Electronics representatives smiled politely. McBride was not the polished advocate that Owens was, but he came across as both knowledgeable and sincere.

"Having attended the Navy's debriefing a few weeks ago, I'm having some difficulty understanding the New England Electronics' protest. The points awarded each contractor in the areas of technical approach and support do not appear to be in dispute. The area in dispute appears to be that of projected cost, so I'd like to have our proposal manager, Charlie Austin, say a few words about that." McBride's written brief had been submitted in rebuttal of the New England Electronics brief, and he knew that a redacted copy of it would be submitted to the competitor. The rebuttal that was about to be made in the presence of New England Electronics had purposely not been elaborated in the written response in the expectation that New England Electronics would not be prepared to respond. He also wanted New England Electronics present for this presentation so that there could be no future claim of Telonics dealing with GAO independently behind closed doors.

"Thanks, Al, and good morning everyone," Austin began. An experienced program and proposal manager, he was an excellent presenter as well, but more casual and less lawyerly in style. "I'm sure that everyone here appreciates the difficulty in estimating the cost of a project of this magnitude. Of course, I don't know what the specific New England Electronics cost estimate is; and I don't know their estimating methodology, but I suspect it is not unlike our own, the industry having settled on fairly common best practices over the years.

"Basically, what we do at Tellonics is to study the customer's requirements and what is to be delivered, and then we break down the work into several high level elements. Each of those elements, in turn, is broken down into lower level

11

elements, and so on into smaller and smaller increments until we have manageable work elements that can be estimated with good accuracy by qualified people who understand the various elements of work to be done.

"In the case of SEANET, we broke down he work into about twenty five hundred independent work elements. This is a big contract, on the order of a billion dollars, so on average each of those individually estimated work packages constitutes almost half a million dollars. We could have created a much deeper work breakdown structure, or WBS as we call it, but our experience has been that if you break down the work into too many small elements, duplications begin to occur. We are quite confident that when we add all these individual estimates up through the cost pyramid, we have a good total estimate for the whole job. Next we apply a number of reasonability tests to the total and its major components. Those reasonability tests look for omissions and duplications and are based on our experience with similar work we've done before.

"Finally, we do what we call an independent parametric estimate for the highest value work packages. For a simple example, consider the weight of a radio transmitter of a certain power and frequency range. Given the specific technical requirements, we can establish with very good accuracy what a given transmitter might weigh. It happens that weight is a very good predictor of the potential cost of the transmitter – the heavier the more costly. And we know how much transmitters that we and others have built weigh and how much they cost, so we can estimate the cost of a new transmitter using predicted weight as the estimating parameter. If you took a look at an old Sears and Roebuck catalog for, say, washing machines, you would note that the more expensive models are relatively heavier. So we add the parametric estimates to the mix as another reasonability check. Then at the end of the day our proposal team gets together with management and we develop our final estimate taking all of these estimating techniques and reasonability checks into account.

"Now, we are all aware that this is not a fixed price contract. The risk is too great for that, and no sensible contractor would offer a fixed price proposal under these circumstances. So in this case the risk is assumed primarily by the Government, which will pay the contractor's actual cost plus a bit of profit depending on the contractor's performance. So the question becomes, how good is the estimate? Statistics on cost estimates for large-scale defense contracts with high-risk development components have found those estimates to be no better than plus or minus twenty percent, and sometimes not even that good. Large overruns are not uncommon, as you know. So the Navy typically sets aside a 10 percent or so reserve to cover that eventuality based on their own cost estimate. They call it a 'should cost' estimate. In other words, how much does the buyer think this work should cost? That Navy should-cost estimate becomes the standard against which all the contractor estimates are compared. Of course, it may not be any better than the contractors' estimates, but at least it's a yardstick for a sanity check. The usual practice of Source Selection Boards, as I understand it, is to treat estimates with skepticism if they differ too much from their own should-cost estimate, but if they are within a ten to fifteen percent range of each other, they are considered valid estimates within the expected range of variability. We do not know the New England Electronics estimate, but if it does fall within, say, ten percent of our estimate, and even if it is lower, they could be awarded the same number of points as we because of the uncertainty of all estimates. That must be the case here, since both New England Electronics and Tellonics were awarded the same 25 points for the cost evaluation element.

"Finally, we all know that in a high risk cost reimbursement contract such as SEANET, at the end of the day it will cost the Navy what it actually costs – not what it was estimated to cost. The award of points for cost estimates, then, has to be based in part on the contractor's past performance. Over the past ten years, Tellonics cost estimates versus actual cost have been well within the expected range, and that fact must have been given considerable weight when arriving at our score. Those statistics were submitted with our proposal and are available for your consideration.

"My point is that none of the three competitors was "the low bidder." There is no low bidder, because these are not bids. They are only estimates, the credibility of which was taken into account by the Source Selection Board when it awarded an equal number of points to New England Electronics and Tellonics. And so, Ms. Surrett, New England Electronics may be the low estimator, but they are not the low bidder, as they claim, and therefore their protest is without merit."

Austin sat down, and McBride concluded the Tellonics remarks. "That's about all we have to offer today, Ms. Surrett. May I suggest that you confer with the Navy's proposal evaluators to confirm that they followed the process that Charlie has just described and that their scoring of the cost estimates was performed as we have suggested? We stand ready to provide any other information that may be of value to you, but beyond that, we await your decision. And thanks again for the opportunity to offer our comments."

McBride's strategy had worked. New England Electronics was not prepared to offer a rebuttal to Charlie Austin's argument that there is no such thing as a "low bidder" in a cost reimbursement contract. That is a simple fact that cannot be reasonably disputed, so it is the evaluator's perception of the reliability of the estimates that is applied in awarding points. GAO now had a peg on which to hang its decision should it decide to dismiss the New England Electronics protest.

Surrett ended the meeting. "Thank you all for coming to Washington today. Your comments have been very helpful to us, and as I mentioned earlier, you'll have a decision within two weeks. Have a safe trip home."

****

Ten days later, on the twenty eighth of June, the GAO decision was published in the Congressional Record. Marcia Surrett placed courtesy calls to inform Richard Owens of New England Electronics and Albert McBride of Tellonics that the SEANET protest had been denied.

Conferring with New England Electronics CEO Ralph Lowry, Owens recommended that no further protest action be taken since it was unlikely to succeed. Together they called Senator Flannery to inform him of their decision.

Flannery's response was predictable. "Not to worry. I'll get those California bastards and that stupid Admiral one way or another down the road. Just wait until they make their first big mistake, and they will make it."

CHAPTER TWO
INVITATION

As United flight 99 touched down at LAX, Jack Decker anticipated a fine August day, the best time of year for southern California, hot, clear, and dry. It would be a relief to leave behind Philadelphia's humidity, particularly oppressive this year. Although eleven years had passed since he had relocated from Los Angeles, he recalled that the Sheraton had an outside pool in the sun, and he looked forward to a few laps.

Arnold Tell's call had come as a nice surprise; they had not talked in a year. "How the hell are you, Jack?"

Decker had recognized the mentor of his early career at once. "Is that you, Arnie?"

"Of course it's me, you old sonovabitch. How come you never call, Jack?" It was typical Arnie Tell, Decker thought. Put the other guy on the defensive right up front.

"Well, I confess I'm not a very reliable correspondent, Arnie, but I'm just fine, thanks. And you?"

"Not too bad for the boss's son, I guess. What the hell are you up to these days, Jack? Still pumping out clones of the last honest work you did? I can see you now with your feet propped up on your desk watching the troops from your second floor window – especially the ladies." Decker was not the womanizer that Arnold Tell had always accused him of being. It was Tell, in fact, who had played that role, although discreetly. Tell was a bachelor, a big man, not bad looking, and the Tell millions (possibly billions) had considerable appeal.

"Hey, Arnie, you know me. I don't work; I watch other people work. I just supervise. Anyway, It's really great to hear from you, but I must admit I'm a bit curious why I'm being called out of the blue by such a super-important person as the CEO of Tellonics." Decker responded as he would have done when they worked side by side. It wasn't boot licking, exactly, but measured stroking to keep the Arnold Tell ego properly lubricated and running smoothly.

Tell made his point at once. "How'd you like to some supervising for me, Jack? I've got a helluvan opportunity that's got your name written all over it."

"You're kidding. I have a pretty comfortable arrangement here, Arnie. No hassles, no problems, at least no big ones. Why should I come back to doing 'honest work,' as you put it?" It was said in jest, but Jack Decker had deeply etched memories of their years together on the Northwind project. Those were exciting years, his best, but they left him with a bitter memory. When he fled to Philadelphia, he promised himself he would never go back to California for more than the essential business trip.

"Look, Jack, this is damned important stuff. I can't discuss it on the phone, but I need you out here. Take a few days off. It's all on me. Come on out and let's talk. If you don't like it, get back on the plane and no hard feelings. Who the hell's gonna even know if you leave that pussycat job for a few days. You could fly to the moon and they'd never even miss you. When are you coming?"

Decker remembered that Tell was a good a closer. "Never ask a question that can be answered with a 'yes' or a 'no'," he had instructed. "You might not like the answer."

"Let me think about it for a day or two, Arnie, and I'll get back to you. I'm really not looking for anything, you know." Decker had wanted to say 'no' right up front, but there had always been a gnawing feeling that running a manufacturing plant was not the best he could do. It was safe, and he did it well, but there was no comparison with the California days.

"The hell you will. I'm calling you. Tomorrow. And you'd better say 'yes' you bloody sonovabitch. Besides, you need some California stuff one more time before you check in to the bloody retirement home." Tell had taken up the habit of using the expletive, bloody, as a substitute for the popular adjective that had colored his language for years.

Decker had agreed to come when Tell called back the next day. There was no harm in just talking, and it would be good to see some of his old associates, especially Arnold; they had been very close friends. It was more a vacation than a job interview, he told himself, and while there were bitter memories lingering in California, perhaps he could put them behind him after all these years.

As United 99 taxied to the gate, the older of the two flight attendants working the first class cabin approached Decker's seat with a smile. Arnold Tell had insisted he fly first class, and now he suspected that he was going to get the full treatment in grand style. "Mister Decker, I've just been informed that there is a limousine waiting for you, and the driver will meet you at the baggage claim area."

The limousine driver's courtesy and attention were exceeded only by her beauty, which was not concealed at all by the navy blue slacks and short "Ike" style jacket, both form fitting. Arnold Tell never missed a trick when he really wanted something, and Jack Decker began to realize that this was to be no ordinary job offer. The silver, stretched Cadillac with smoked windows was of the type usually reserved for weddings, funerals, politicians and rock stars, and Decker felt as though the entire airport had stopped for a look as he entered the door held open with style and grace.

"I've never felt so self-conscious in my life," he admitted honestly to the driver as the limousine pulled away from the curb to join the bustling array of cars, shuttle vans, taxis, and busses that populated, like blood cells, the veins and arteries of the great horseshoe of Los Angeles International Airport.

"Oh, but you've got to admit it's kinda fun to live like a movie star once in a while, Mr. Decker. My name is Cindy, by the way. Mr. Tell said the limo is at your disposal while you're in L.A., so you might as well get used to it."

"Terrific," Decker muttered. He must not let this luxury and attention go to his head. If he knew Arnold Tell, and he did - very well - the whole scenario was fully orchestrated, so he would have to be on guard through what he now began to see as a threatening situation. Despite their old friendship, Arnold Tell would not go to all this trouble and expense, and personally, unless he had a very serious problem.

"Mr. Tell said to tell you that he could see you about three o'clock, so I'll take you to your hotel to check in and have some lunch. He said he's sorry he couldn't meet you for lunch today, but don't make any plans for dinner." Flight 99 had left Philadelphia at nine and landed at LAX at noon -- six hours in the air considering the three-hour time zone change, so its westbound passengers would have a long and tiring day. By the time he saw Arnold Tell at three o'clock, his own biological clock would be at six. Tell had set the stage well, Decker concluded.

The L.A. freeways were as he remembered them, but more crowded. Traveling north on the San Diego then east on the Ventura, they arrived at Universal City in less than 30 minutes. Cynthia Robbins was a skilled driver, maneuvering the dachshund-like, stretched Cadillac cautiously but systematically from lane to lane to take advantage of openings in up to five parallel streams of traffic. Her conversation was informative as she revealed that she had been one of Arnold Tell's three personal drivers for the last four years. The blonde ponytail concealed her thirty-two years well. An obvious "hard body," as her gym-mates would testify, she was a trained bodyguard, and armed. Decker wondered where she could possibly be concealing a weapon under the form-fitting uniform. The other two drivers were men, usually working the night hours in Los Angeles, and accompanying Tell when he traveled. Since men like Arnold Tell were prime targets for kidnappers, terrorists, malcontents, and the mentally disturbed, personal security for the rich and famous had become a thriving new profession of which Cynthia Robbins was a part; but her enthusiasm made it clear that this was more to her than just a job. There was great personal loyalty to Arnold Tell reflected in her remarks, and Decker wondered if their relationship was purely professional. Knowing Arnold, probably not.

Decker told her a guarded minimum about his previous association with Arnold Tell. They had been classmates at Cal Tech in the seventies. Tell was four years older and had entered college after serving in the Marines during the Vietnam war. Decker had been awed by the swashbuckling Tell, the returning warrior. However, Decker was the better student, spending the next four years tutoring Tell in matters academic, while in turn, being tutored by Tell in matters practical (and romantic -- also not revealed). When they graduated, Tell insisted that Decker work with him at the family firm, Tell Electronics in Burbank, the company subsequently incorporated and renamed Tellonics. They had been a terrific team at Tech, why not in the real world? Beyond that, Cindy was told only that Robert Arnold Tell the elder had insisted that his son start at the bottom (albeit as an engineer), and that they had worked together off and on for almost twenty years before he had accepted a position to establish a new manufacturing facility for Delaware Microwave in Philadelphia. The circumstances surrounding his leaving Tellonics were not forthcoming.

They arrived at the Sheraton Universal at twelve forty five, and Robbins arranged to pick up Decker in two hours for the short drive to Tellonics' Burbank headquarters. Although he had visited the Sheraton-U on many business and social occasions, he had never been a guest there and was unfamiliar with its facilities, so he was pleased to learn as he checked in that there was, indeed, an outside pool as well as one inside the hotel along with all the other fashionable physical fitness facilities: gymnasium, sauna, and even a private jogging trail. An executive suite had been reserved for him as a guest of Tellonics, all services of the hotel were available to him without charge, and he was not to hesitate to ask for anything he might desire -- without exception, the desk clerk pointed out with a knowing smile. Arnold Tell had left no stone unturned.

Jack Decker was no novice at negotiations, and he knew that he must be at his best to deal with Arnold Tell. Friendship aside, these two would use every skill to make the best of a situation, but Tell would go further than Decker, having a somewhat lower ethical threshold, and Decker knew it.

After unpacking in his penthouse suite, a large, three-room arrangement with a magnificent view of the San Fernando Valley, he dressed for the pool and went there immediately. After a few laps and a poolside lunch of only chef's salad and orange juice, he stretched out on a sunlit lounge, giving instructions to be awakened at two o'clock. He didn't really sleep, but dozed comfortably in the warm California sunshine of a beautiful, dry, eighty-five-degree day.

Jack Decker's thoughts drifted to his departure from Los Angeles in 1990 and the events leading up to his break with the past to start a new life. His wife Martha's sudden illness and death had been devastating. Still slender and athletic at 40, she was the prototypical California sun lover, perpetually tanned and healthy. Then the sudden lack of energy and weight loss, originally thought to be anemia, was attributed at last to leukemia of a type not well understood at that time. She was gone in three months after undergoing painful treatment that proved of no value. It was the darkest episode of Jack Decker's life, for he was truly devoted to this remarkable woman -- lover, companion, advisor, and mother of his three children. Unable to deal with the constant reminders of their life together, he had retreated to Philadelphia, a place that he and the children had come to regard as home in the last eleven years. He wondered if returning to the surroundings in which he and Marty had spent their years together would bring back all of the memories that had forced him to seek refuge.

"It's two o'clock, Mr. Decker," the pool attendant's pleasant voice summoned. Returning to the present, Jack

Decker felt a rush of apprehension. In just one hour he would be walking into Arnold Tell's office for the first time in more than a decade.

CHAPTER THREE
CLASSMATES

The upbringing of Robert Arnold Tell, Junior, was not a routine endeavor. Big, strong, athletic, intelligent, with a lust for life and adventure, he would have been admired by adults and popular with peers even if not the son of a multi-millionaire. Tell the senior was commonly called Robert, Tell the junior, Arnold, to avoid confusion. The elder Tell's closest friends still called him "Sparks," an artifact of his Navy days as a wireless operator aboard USS Pennsylvania during the Great War. The entrepreneurial founder of Tell Electronics, Robert was persistent in his effort to balance Arnold's upbringing with a first-rate education. His mother Margaret's influence contributed a life-long sensitivity to others, especially the needs of the less fortunate. The combination of these inherited and learned qualities produced a complex individual -- fun loving and aggressive on the outside, thoughtful and sensitive on the inside. The heir to the Tell industrial empire had been deliberately and carefully sculpted to follow in his father's footsteps.

As Tell junior approached his eighteenth birthday, Tell senior expected his son to continue his education uninterrupted following his high school graduation. With a strong academic record, and with his father's encouragement, he had applied to and been accepted by the University of California at Berkeley, Stanford University, and California Institute of Technology. But Tell junior, with his appetite for adventure, chose to follow more closely than intended in his father's footsteps. On his eighteenth birthday, September 4, 1965, parental consent no longer required, Arnold Tell enlisted for four years in the United States Marine Corps.

The parents were both shocked and upset by this unexpected but irreversible event. Arnold's explanation was simply that he wanted to serve his country, as his father had done, and to see another side of life. It was only four years, after all, and there would be plenty of time after that for college and a career. Tell senior was not so much disappointed as he was concerned for his son's safety. Two months earlier, in July of 1965, the United States had sent its first military advisors, as they were then called, to a dangerous corner of the globe, a mysterious place carved out of what had previously been colonized as French Indo-China, a place now called Viet Nam.

Arnold Tell endured boot camp better than most at San Diego's Marine Corps Recruit Depot, MCRD. Men, large boys actually, from all walks of life were thrown together in a melting pot of equality such that none was distinguishable from any other. Ordered about from dawn to dusk, berated, humiliated, driven to extremes of tortuous physical endurance, there began to coalesce a well ordered cadre of men whose mutual experience led to a camaraderie that held them together as one. At the end of thirteen weeks they were fit to be Marines, every one a qualified rifleman, and conditioned to accept orders without question. Each was dispatched for further training to an operational unit best suited to his individual qualifications. Tell's excellent vocational test scores qualified him as a candidate for electronics training, and he was ordered to the U.S. Navy's electronics training school at Great Lakes Naval Training Center near Chicago. After six months of electronics training with the sailors and two other Marines, he returned to MCRD San Diego for an additional three months of specialized training in radar.

Tell completed two yearlong tours in Viet Nam during his four-year enlistment, rising to the rank of Technical Sergeant, a five-striper, in less than four years, a portent of the extraordinary leadership qualities that would take him to the heights of corporate achievement. As a radar specialist, he had been trained to repair, set up, tear down, and relocate, as the ebb and flow of battle required, mobile radar equipment used for artillery fire control or to coordinate ground-air operations. As it happened, Viet Nam was not that kind of war most of the time, so Arnold Tell the rifleman was called to that fundamental duty of every Marine on many occasions. He participated in firefights and search-and-destroy missions for which the Viet Nam conflict became infamous. When his four-year enlistment expired in August of 1969, Tell was not compelled to reenlist even though the Viet Nam conflict continued, but he was offered an opportunity to remain in the Marine Corps and be sent to Quantico, Virginia, for training as a commissioned officer. Having observed the mortality rate of Marine Corps second lieutenants first hand, he respectfully declined. In later years, he refused to discuss his Viet Nam experience.

Arnold enrolled at the California Institute of Technology in September of 1969, his prior acceptance four years earlier still acknowledged. At first he was shocked by the virulence of the anti-war sentiment. Eventually he came to acknowledge that there were valid arguments against U.S. involvement in Viet Nam and some of the tactics being used, especially the defoliation and bombing. Nevertheless, on balance he believed that U.S. involvement was justified – the domino effect rationale -- and he did not hesitate to say so when confronted by those of another viewpoint. On occasion the arguments nearly came to blows, but Arnold's stature limited his adversaries' abuse to verbal. A charismatic and effective spokesman for an unpopular position, it was inevitable that he would attract others of like sentiment, and one such was fellow freshman Jack Decker.

Jack Decker was raised in southern New Jersey, just across the Delaware River from Philadelphia. His father, Charles, was a native Californian and a World War Two army veteran. After his return from Germany and discharge from the Army, Charles Decker took advantage, as did many, of the GI Bill of Rights to continue his education. The Veteran's Administration paid his tuition at the California Institute of

Technology for four years, and he graduated cum laude with a Bachelor of Science degree in Electrical Engineering in 1952.

Jack had been born a year earlier to Charlotte and Charles, high school sweethearts, who were married when Charles was a Cal-Tech sophomore. Like many couples starting a new life after the war years, Charlotte's salary as a secretary supplemented the Veterans Administration's small stipend to provide a modest but adequate income. The baby's arrival, however, caused them to use most of their savings since Charlotte was unable to work full time. Therefore, at graduation it was essential that Charles find employment immediately. He accepted a position with RCA in Van Nuys, California, where he worked for the next seven years and was recognized as a high potential employee. When Jack was eight, his father was transferred to RCA headquarters in Camden, New Jersey, home of the old Victor Talking Machine Company now owned by RCA. The Decker family remained in the Southern New Jersey area for the remainder of Charles' career until his retirement in 1985.

Charles had not insisted that Jack become an engineer, but the boy showed natural ability in science and mathematics. Like his father, Jack became an enthusiastic Radio Amateur, and when it came time for a career decision during Jack's senior year at high school, Charles' alma mater, Cal Tech, was the obvious place for son Jack to earn an engineering degree. The year was 1969, and the Viet Nam War was heating up. Jack considered enlisting, but relying on his parents' advice, he would just take his chances with the draft. If selected, he would serve.

With their very different backgrounds, it was unlikely that the paths of Jack Decker and Arnold Tell would merge into one, but that evolution did occur and was of lifetime consequence to both.

CHAPTER FOUR
REUNION

Arnold Tell was in serious trouble. As Chairman of the Board, President, and Chief Executive Officer of Tellonics Corporation, he was one of the most powerful men in the aerospace industry, but he did not have the luxury of complete control enjoyed by his father. The company having gone public ten years ago after Robert Arnold Tell the senior's retirement, stockholders were demanding increased dividends and divestiture of unprofitable operations, while Arnold Tell's board was pursuing a long-term growth and investment strategy at the expense of today's dividends. Mason Crenshaw, espousing the coupon clipping stockholders' viewpoint, was at Tell's heels in pursuit of the CEO position and might soon have the votes to swing it. And then there was the matter of the SEANET program.

Despite a formal protest by New England Electronics, the contract for developing the Space Environment Automatic Naval Electronic Telecommunications system had been awarded to Tellonics' Defense Systems Division two years previously, in March of 1998, after five years of fierce competition for the largest communications contract ever awarded by the Department of Defense. Now the program was months behind schedule, cost overruns were accumulating to the point of embarrassment to both the company and the Navy, and Rear Admiral Russell "Rusty" Sullivan, the Navy's program manager, was demanding that Tellonics take immediate steps to bring the project under control before congress and the press got wind of the situation. It was no longer a problem that could be contained by the Navy and its prime contractor, DSD, the Defense Systems Division of Tellonics, and direct intervention by Tellonics Corporate Headquarters was demanded.

Lawrence Hamilton, a Tellonics Vice-President and General Manager of the Defense Systems Division, was a competent executive. He had run the division profitably for the last two years. Since the department from which it had its beginnings was established as an independent operating unit, DSD had quadrupled in size and met its business objectives every year without exception. However, SEANET had proven to be a very large bite for DSD to swallow.

The size and complexity of the project had required Defense Systems Division to increase its staff by a third in just six months time, largely through recruitment of the competition's personnel -- typical of the movement of skilled specialists who follow the contracts du jour in the aerospace community – and to a lesser extent by the employment of temporary personnel. Physical plant and equipment had grown proportionately, and large capital investments had been required to provide the specialized research and development and manufacturing facilities needed for SEANET's design and production. It was an ambitious and complex undertaking, and it was inevitable that things would not go as smoothly as those who had planned it had assumed with characteristic optimism. They never did. "All of Murphy's laws," Larry Hamilton opined, "were discovered by the ancient Egyptians during the construction of the pyramids, the first major project in antiquity. When preparing proposals they are invariably dismissed, but they are always rediscovered during the first six months of a new project. The most important of these laws is: whatever can go wrong, will. The second is: when something does go wrong, it will do so at the worst possible time."

Charles Austin, the manager of the successful proposal team that had brought in the contract, had been appointed the original SEANET Program Director as a reward; but unaccustomed to managing projects of such a large scale, he was found wanting in that role. Hamilton had replaced program directors twice in the two years since the contract had been booked, each time with the most promising candidate available; but expectations of turning the project around were not met on either occasion, and Larry Hamilton was at a loss as SEANET continued to miss important milestones and exceed budgets.

Arnold Tell had learned most of what he knew about running a business from his father, and one of those lessons was that the most important function of an executive is the selection of competent subordinates. "I've probably spent more troubled days and sleepless nights worrying about who to put in this job or that, than I have spent on all of my inventions," the elder Tell would lecture. "Making a good choice will keep you out of trouble; making a great choice once in a while will pay off handsomely; but making a bad choice can be a disaster." So Arnold Tell had relieved Larry Hamilton of this SEANET management decision by taking the matter into his own hands. He was going to get the best man he could envision for the SEANET job, Jack Decker.

It was Decker who had been responsible in large measure for Arnold Tell's corporate success. When the old Tell Electronics Company had been awarded the Northwind Project in the mid-eighties, it was not unlike the present situation. The Northwind program had been the straw that almost broke the camel's back, making the company in its totality an unmanageable organization; so the company had been reorganized into six separate divisions, each subsidiary an independent operating unit. The mother company, suddenly booking over a billion dollars in annual sales, had been renamed Tellonics Corporation and had gone public to raise badly needed capital for expansion. Arnold Tell, running the company after father Robert's retirement, had brought Decker over to DSD as Program Director of Northwind, and the contract became, ultimately, an enormous success -- completed on time and profitably for a well-satisfied customer, the United States Air Force.

Both Tell and Decker had continued to prosper until that unexpected and tragic event changed the course of their association. Martha Decker's untimely death in 1989 had led to Jack Decker's departure for Philadelphia within six months. Now Arnold Tell was determined to get him back.

****

"Mr. Decker has arrived at reception, Mr. Tell," Susan Anders announced on the intercom with uncommon formality.

"Will you pick him up please, Susan?" he requested politely of his personal assistant and secretary. She had been his secretary and administrative assistant during her fifteen years with Tellonics, moving up with him to each successively higher position: Associate General Manager of Tell Electronics, then President of Tellonics, and finally Chairman and Chief Executive Officer. She was a charming, intelligent, and attractive woman who had made it clear to Arnie Tell at the outset that theirs was to be a strictly professional relationship. She was also of the new breed of executive secretary -- confident and assertive, capable of taking most of the heavy administrative load off of her boss's shoulders while acting as an unofficial chief of staff by insuring that his instructions to subordinates were communicated clearly and acted upon promptly. A polite call from Susan Anders was tantamount to a demanding call from the Chairman, himself, and she was treated with considerable deference. The unusual act of having her personally meet Jack Decker at the security desk was a symbolic gesture that Tell knew Decker would recognize as attaching great importance to his visit.

Cindy Robbins, maneuvering the limousine deftly through the rectangular grid of San Fernando Valley surface streets, had transported Decker from the Sheraton to Tellonics' Burbank headquarters in just ten minutes. With a pleasant smile she indicated that she or one of the other drivers would pick him up when he concluded his business for the day.

"Jack Decker! What a pleasure to see you again," Susan Anders greeted him. They had known each other since she became Arnold Tell's secretary, five years before he had left for Philadelphia, and they had enjoyed pleasant, informal chats on numerous occasions. In Susan's view, Jack was a really nice guy, and she liked him.

Decker was flattered, as intended. "The pleasure is all mine, Susan; what a very nice surprise. You look terrific. How have you been?" He grasped her hand as she extended it, and they both felt the warmth of greeting an old friend.

He signed in with the guard-receptionist, and Anders signed as his escort while he donned a visitor's identification badge, then they departed for the executive penthouse on the twentieth floor. As they left the reception desk, he looked in the display case for the honored antique Model 17 radio, Robert Tell's first patented invention, and he was comforted that it still resided there after all these years.

Jack and Susan exchanged a few comments about themselves, barely touching the surface of all that had transpired since they had seen each other last, when suddenly the elevator discharged them into the high-tech atmosphere of the executive reception area leading to Arnold Tell's suite of offices.

"Jack! It's about time! Come on in, you old sonovabitch," Tell beamed, making no attempt to filter his colorful language in Susan Anders' presence. He was genuinely delighted to see Decker again. Despite their frequent differences over the years, they had been classmates, colleagues, and personal friends for a quarter century.

"Hi, Arnie. It's great to see you again." Decker was pleased also. They had been a terrific team, and he had a warm affection for this sensitive, if bombastic, man. He had always felt protective over Tell since their days at Cal Tech when he had been his confidant, conscience and tutor. "You look just like you did last time -- except about ten years older."

"Screw you, Decker. I'll whip your ass any time, just like I always could."

"Well, here we are, going at it again. Just like old Times," Anders chimed in. She was perfectly comfortable in this situation, not just witness but participant in the occasion of two old friends meeting after years of separation. "I'll have to leave you two guys, now -- somebody has to do some useful work around here. See you in a little while, Jack." She departed gracefully, leaving the two to share the moment alone and then to conduct their business privately.

They talked for half an hour about their personal as well as their business lives, occasionally reminiscing or laughing aloud about some incident of their early years together. Decker had settled in to an important but not particularly challenging position, content to raise his kids, get them through college, and eventually married. He said little of his association with women over the years, responding to Arnie's question about remarriage with, "Well, I just never met anybody else who could replace Marty."

Tell related his personal experiences with the observation that he had had enough romance for the two of them and was de-lighted to have filled in for him. Concerning business, he was obviously proud of his and his company's achievements. DSD and the total company had done very well indeed under his leadership. Despite the fact that he was the son of founder Robert A. Tell, Sr., it had not always been easy to move up the corporate ladder. As he put it, he had been obliged to "... keep my foot in the face of the guy below, and my nose in the ass of the guy above." He confided in his old friend his concerns about Mason Crenshaw's politicking for a shift in board membership at the next annual meeting.

"Who is this guy, Crenshaw?" Decker inquired.

"Well, actually, he's a pretty good troop. I'd offer him a top job anytime. The only thing I've got against the sonovabitch is that he's after my job." Tell explained that Mason Crenshaw was chief executive of Commware, Inc., a small but excellent communi-cations software firm with which several Tellonics divisions did business. Crenshaw had been appointed a member of the

Tellonics board of directors about three years ago, despite the potential conflict of interest as a supplier to the company, because he was extremely capable and brought much needed technical balance to the boardroom. Since that time, Crenshaw had expressed an unexpectedly narrow, short-term viewpoint and had been cultivating stockholders, actively seeking their proxies in an attempt to displace Tell as Chief Executive. He proposed to maximize current profitability at the expense of long-term growth, while Tell and the board majority took the opposite view, the entrepreneurial spirit of R. A. Tell, Sr. surviving in the bones of R. A. Tell, Jr. Crenshaw's plans had significant appeal to the multitude of shareholders and to the more conservative members of the board. Few but the institutional investors were interested in future potential, unwilling to milk the corporate cow for a meal today at the expense of tomorrow's appetites.

Despite the large number of inherited shares still held by Arnold Tell, it was going to be a tough contest. It was taking a lot of Tell's time to fight off the challenge, and he was distracted frequently from the important business of running the corporation. That steered the conversation to the SEANET program, which had to be brought under control in order to counter Crenshaw's certain charges of mismanagement if the matter attracted too much attention.

"What do you know about SEANET, Jack?" Tell inquired.

"Not too much, Arnie. I've heard about it of course, and I know it was a great surprise when you took it away from New England Electronics. So how's it going?" Decker asked.

"In a word, shitty," Tell complained. "Nobody seems to know how to run a bloody program any more, Jack. SEANET is a program manager's dream. It's got everything, and it's going to be a tremendous system if we can ever get the damn thing built. Best customer in the world, the Navy, state of the art technology, a real engineering challenge, and good people working the job. And the Navy needs it, Jack. All I need is somebody who knows what the fuck he's doing to run the sonovabitch." The word, bloody, didn't seem to fit.

"Is that why I'm here, Arnie?"

"Well, the thought crossed my mind," Tell smiled. "Interested?"

"Not a chance, Arnie. What the hell do I want to break my butt for, when I've got a nice, comfortable job in Philly?"

Tell was waiting for that expected response, and he had a well-prepared argument to present to Decker; he delivered it with his accustomed persuasive skill.

"Let me tell you about the system," he began, knowing that Decker, like his father, loved a good technical challenge more than anything else. He explained that the U.S. Navy's ability to communicate had, historically, given it the tactical edge over its adversaries, but that the edge was gone. Realizing that it had lost the technological lead in this area, top thinkers in the Pentagon had, with industry's help, begun searching for an answer over ten years ago. Tellonics' DSD had participated enthusiastically in the early studies and research projects which led ultimately to the SEANET program. Tell had invested a significant portion of Tellonics' internal research and development funds in related technology and had shared expenses with the Navy by accepting fixed price study contracts at half their actual cost. As a result of its investments, Tellonics had been well positioned to compete for the actual development and initial production contract that had been awarded after years of work by both Government and industry.

When SEANET was completed, Tell explained, any shore station, ship, submarine, or aircraft in the fleet would be able to communicate directly with any other anywhere in the world. As far as the user was concerned, the network would be so transparent that it would be like placing a phone call across town. SEANET would have practically unlimited capacity and would be able to handle video, audio, code, and data transmissions either in the clear or with security encryption. It would be possible for naval commanders to send instructions simultaneously to selected units or commands, or for that matter, to all of them -- anywhere, any time.

The heart of the system would be its communications satellites positioned around the earth in synchronous orbits such that at least two were always available to any spot on the globe or its atmosphere. The ultimate cost of the system would be enormous because, in addition to the expensive satellites at over two hundred million dollars each, much of the communications equipment currently in use by the Navy would have to be modified or replaced.

Tell explained that SEANET was heavily dependent on some proprietary trade secrets developed under one of DSD's internally funded research and development projects. A former associate of Decker, Dr. Milton Karinski, and his team had found important new ways to pack information into narrow increments of the electromagnetic spectrum such that an extremely large number of channels could be established to transmit simultaneous messages. At first it had been thought that Karinski's concepts were impractical, but it was shown by computer simulation that they could work. Limited experiments were conducted to prove the fundamental scientific principles, then some prototype hardware was constructed and computer programs written for a practical demonstration. The results of those demonstrations convinced the Navy that a solution to its worldwide communications problems might be found in those concepts, and so began the joint military and industrial research that led four years later to the competition for the SEANET contract.

Having won that competition, Defense Systems Division of Tellonics was under contract to the U. S. Navy's Space and Naval Warfare Systems Command, SPAWAR, to design, build and test the first three SEANET satellites and a quantity of equipment sufficient to equip three shore installations, twenty-five surface vessels, two submarines, and one hundred aircraft. The value of the contract was slightly over one billion dollars. It had been a plum sought after by five industrial joint venture teams representing almost every military electronics contractor of substance in the country. Depending on the success of this first phase, it was the Navy's intent to negotiate a low rate initial production, or LRIP, contract with the developer, then to conduct a series of competitions for the production of the remaining equipment needed to install SEANET worldwide. Tell pointed out that, as the incumbent of the development phase, Tellonics and its partners would have a competitive advantage when profitable production contracts followed.

SEANET was to be deployed as a total "weapon system" -- aerospace jargon descriptive of the project even though no weaponry as such was being procured. DSD's role in this vast undertaking was that of prime system contractor. While some of the hardware and software was to be provided by DSD and her sister Tellonics divisions, most of it was to be provided by major subcontractors -- one of which was Mason Crenshaw's Commware -- through a complex, multi-layer tiering of further subcontracting and purchasing. In all more than a thousand companies -- as well as one could estimate -- would provide hardware, software or services for SEANET. It was DSD's responsibility to specify, procure, integrate, and deploy the system, placing it into operation and turning the key over to the Navy when it was done. Thus it was termed, in the language of the business, a "turnkey" operation. It included not just the equipment, but all of the support elements needed to place such a system in operation and keep it running: spare parts, support equipment, technical data, training, and support personnel. Tellonics, as prime contractor, was responsible for making the whole system work.

The first of the three satellites was to be launched in December of 2000, just months away; and the other two in February and April of 2001, respectively. By that time the three shore stations, ten ships, twenty five aircraft, and one submarine were to be outfitted with their companion equipment to support comprehensive testing by the Navy's Operational Test and Evaluation Force, OPTEVFOR, scheduled to begin in May of 2001. The remainder would be outfitted as additional equipment became available.

The schedule was important for several reasons. Firstly, the sooner the better -- the System was needed badly. Secondly, there were budgetary constraints. If programmed funding was not expended in relatively close agreement with the planned funding profile, there was always a chance of losing unexpended funds to other government programs. Thirdly, being late would be an embarrassment to both the company and the Navy. Ships had to be scheduled years in advance for their overhauls or restricted availability periods in port for equipment installation. Finally, and most importantly, time cost money. The longer a program lasted the more it would cost.

SEANET was at the high-risk end of the procurement spectrum and was a cost incentive contract. DSD would be paid all of its actual costs. But a target cost was established at the outset, and DSD's profit would depend on how well the target was met: below target cost and a greater fee would be paid; right on target cost the target fee would be paid; over target and the fee was reduced drastically. In fact, if the actual costs exceeded the target by twenty five percent or more, DSD would receive no profit at all and would recover only its actual costs. There was also the threat of contract termination that could ruin a contractor like DSD in which major investments had been made to support a single contract. Later phases of the program would, no doubt, be contracted on a negotiated or competitive fixed price basis once the original development risk had been overcome, but for now it was cost reimbursement with all its prospective risks and difficulties.

It was ironic, Decker thought to himself, that the public at large believed that defense contractors got rich at the expense of the taxpayer on such contracts, when in fact, they averaged considerably less after tax profit than they could have made by investing the same amount in treasury bonds.

Decker wanted to know how SEANET was progressing with regard to cost and schedule. Tell would not deceive him; he'd find out soon enough anyway. Besides, Arnold Tell was accustomed to leveling with his subordinates, and he expected them to do the same with him no matter how bad the news. One of Tell's favorite expressions was, "bad news does not improve with age." So the bad news came next in Tell's presentation.

"Not too badly for a CPIF contract. We're about four months late at midpoint of a three year first article delivery with some pretty good workarounds still available. Costs are projected at about fifteen percent over target, but I think we'll do a little better than that if we can pick up the schedule." Even bad news could be presented in a positive way. "The biggest problem is the bloody Navy. Every civil service engineer on the Navy's payroll pushes us to get his favorite little goodie into the system whether it's in the bloody contract or not. Then the bastards are in our pants counting every bloody penny we spend and want a fifty-page variance report every time we're a day late or a dollar over. I'd like to shoot the sonovabitch who invented that bloody CSSR shit." Tell was referring to the Navy's Cost/Schedule Status Reporting system in which cost reimbursement contractors were required to share their most intimate and detailed cost and schedule performance data on a continuing basis. "Keeps us honest, I guess, but if that bloody Admiral calls me one more time this week on some nit-picking, ten dollar overrun, I'm gonna put out a contract of my own on the red-headed sonovabitch."

Decker was surprised to hear that Arnie Tell was so intimately involved in the program. It was unlike him to get involved in the details of an individual project. Like his father, he was a delegator, a man who trusted his subordinates to take care of business. If they couldn't, he'd get somebody who could – as he was doing now. As a consequence of this knowledge of Tell's management style, Decker inquired how Larry Hamilton was holding up under the pressure.

"Larry's doing fine, under the circumstances. I've arranged for you to spend the day with him and his people tomorrow so you can get a first hand look at things over at DSD. But he can't do everything himself. He's appointed a couple of pretty good program managers to run SEANET, but none of them has been able to get it together. We still have a couple of ideas for him to try, but basically, this looks like a job for super-Decker."

"Arnie, I'd be lying if I told you I didn't miss the action. But Christ! This is definitely a career limiting assignment. I'm not sure anybody can pull this off if its as bad as you say, or for that matter that it can even be done at all. You want me to quit a nice job with multi-perks and put my ass in your sling?"

Tell was ready with the WIIFM, or "whif-um" as he pronounced it, the letters standing for "what's in it for me?" He always included plenty of "whif-ums" when he was making a pitch, and this was no exception. This might be the most important sales talk he had ever given, and he was prepared to make Decker an extraordinary offer. "Hey, Jack! You know I'll take care of you 'ol buddy."

"Could you be just a little more specific, Arnie?" Decker prodded.

"Well, for starters a full corporate Vice Presidency, and you'll report directly to me. Of course, since you'll be over at DSD physically and the project is part of their profit center, you'll have to get along with Larry."

"No problem. At least not for me," Decker interjected. He was surprised that he had said it. This was no time to offer encouragement. Tell caught it too.

"And you'll be the main man on SEANET. No questions. No interference. Just get that bloody Admiral off my back. I'll take care of Crenshaw. You take the Admiral."

"Do I get paid, Arnie?"

"Oh, sure. I forgot. Twenty five percent over your current salary. We'll pick up any losses from your entitlements at Delaware Microwave and anything you lose in the move to California. If you make money, you can keep it. All the V.P. perks, of course -- stock options, executive bonus plan, company car, and all that stuff. And I think you're entitled to full reinstatement of all your longevity benefits from your last hitch with Tell Electronics after a couple of years. I'll have to check with HR on that one."

"So far, I'd be just about breaking even, Arnie. It's expensive in California."

"Tell you what, Jackie boy. You give me your answer by the end of the day tomorrow, and I'll throw in a year's pay up front and pick up the taxes." Tell thought this would do it. He was mistaken.

"How about an employment contract, Arnie? Two years guaranteed at least, even if I screw up and get fired the first day on the job." This was something Decker didn't really care much about, but he had to ask for something more than Arnie Tell was offering just to set the proper tone right up front.

"What have you been doing for the last ten years, running a bloody factory or selling used cars? O.K. you bloody thief. Done. Just give me an answer tomorrow and no screwing around for two weeks." The negotiation was completed. Now it was up to DSD.

Arnie Tell pressed an intercom button. "Susan, dear, will you come escort this sonovabitch out of my office before he bankrupts us?"

They said their goodbyes, and Tell reminded him that the limo would pick him up at seven thirty for dinner. Casual dress.

They had talked for an hour and a half and covered a lot of ground. Decker was very tired after his long day that had started at four in the morning in Pennsylvania. Now it was almost eight Eastern Time, five Pacific, with a long dinner to go. "Thanks a lot, Arnie, but if you don't mind I'll just catch something at the Sheraton, watch the news, and get some sleep. Let's have dinner tomorrow, if that's o.k."

"Jesus Christ. Five hours in town and you've got something lined up already. You're not making it with my bodyguard are you?" They both had a good laugh. "Then the limo will pick you up at seven in the morning for your DSD trip. I'll see you tomorrow night."

Decker was pleased to be with Susan Anders again, and she with him as they took the elevator down to reception. "How'd it go, Jack?"

Decker replied, "He's a pretty convincing guy, that boss of yours. I don't know, I've got to sleep on it and get a first hand look tomorrow. I promised him a decision one way or the other tomorrow night. Right now, I'd say it's a toss up.

CHAPTER FIVE
SEANET

Jack Decker's biological alarm clock woke him at four in the morning since it was seven in Philadelphia. He usually had no jetlag problem when traveling from east to west since he would fall asleep early and awaken early. The next hours were a half-awake, half-asleep oscillation during which he alternately thought then dreamed about the previous day's events, terminating finally with the ring of his six o'clock wakeup call. He was completely refreshed after he shaved and showered then dressed for the day. A call to room service for coffee had been placed the night before to initiate the final phase of his standard east-west travel routine, and copies of the Wall Street Journal and the Los Angeles Times were delivered with the coffee as he had requested. He had fifteen minutes with the coffee and papers before leaving his room at five 'till seven, just enough time to capture the flavor of yesterday's world from a scanning of the headlines and reading a few key paragraphs. He placed the papers carefully on the desk so they would be there when he returned at the end of the day for a more thorough reading.

"Private Decker reporting for duty, ma'am," was his greeting to Cindy Robbins as she opened the door of the silver limousine just outside the main lobby door.

"Good morning, Mr. Decker," she replied with a cheerful smile, glad to see him in a good mood. She had liked Jack Decker immediately and was pleased to confirm that he had a sense of humor as she had expected he would. Her instincts about people were excellent, an essential requirement for the security business. He liked her too. It was a father-daughter kind of relationship with which they both felt comfortable, although he could not help but notice again her marvelous figure in a decidedly non-fatherly way. She found him attractive as well.

Tellonics' Defense Systems Division was located in the city of Buena Park in Orange County, about twenty miles south of Los Angeles. Originally housed in facilities that had evolved over the years on the old Tell Ranch in Burbank, there was no further room for expansion when Tellonics was reorganized in 1982. Buena Park had been selected for the new DSD facilities primarily because of the abundance of aerospace talent available in Orange County. While almost 500 key employees had been relocated at company expense when the first of the new facilities was opened there in August of 1982, most of the staff was recruited from the vicinity of the new location. Jack Decker had worked there for almost seven years before leaving for Philadelphia, and he was anxious to see the familiar surroundings and some of his old friends.

The trip to Buena Park would take about thirty minutes normally, but during the early morning rush it could take much longer. This was a particularly trying day, and even though Jack and Cindy chatted on and off during the trip, she gave primary attention to the stop and go traffic. Decker also had a more complete scan of the Wall Street Journal, which Robbins had carefully laid out on the forward seat of the limo. They arrived at the main gate of Defense Systems Division at six minutes before eight. The silver stretch caddy, recognized at once by the security guards, was waved through without a stop, as it always was, bypassing the main stream of traffic being checked in one car at a time.

"Here we are, Mr. Decker," announced Cindy as she pulled up directly in front of the executive office building, one of a complex of six major structures spread over eighteen acres and containing over five hundred thousand square feet of floor space. Decker was uncomfortable each time Robbins exited her driving position to open the curbside rear door for him, but he went along with the formal protocol even though it violated the rules of his chivalrous upbringing. "I'll be around all day if you need the car. Just let the receptionist know, and she can reach me by radio."

"Thanks, Cindy. I'm not sure what the agenda will be, so I'll see you when I see you. Have a good day," responded Decker as he turned with a warm smile and headed for the "brass and glass" building, a nickname the executive offices had acquired when the structure, the last in the complex, was completed.

Obviously expected, he was ushered quickly through reception, and a uniformed guard escorted Decker directly to the executive conference room. It was a room used exclusively for important internal meetings -- usually those attended by the Division General Manager or for VIP visitors -- so it was obvious that his visit was being treated as an important event at DSD. There he was greeted by an array of new faces as well as several familiar ones. His good friend Larry Hamilton was present, obviously in charge, and Doctor Milt Karinski, with whom he had worked on several occasions, was present as well. Coffee and Danish were served -- Decker stuck to the coffee -- during several minutes of greeting the old and meeting the new, when, at exactly eight fifteen, Larry Hamilton politely directed, "Well, shall we get started, Gentlemen and Ladies?"

Hamilton moved casually to the podium at the front of the room, as others took seats at the enormous, rectangular, walnut conference table, its length perpendicular to the front wall and arranged so that those seated could look down the center of the table with a clear view of the speaker and projection screen. They took seats in accordance with an unspoken pecking order, not prearranged but worked out mentally by those present on each occasion. At the farthest end of the table, facing forward was the seat reserved for the General Manager, Larry Hamilton, with the cultural understanding that, if Hamilton were not present in a particular meeting, the holy seat would remain empty. If one were among the higher-ranking present, then one would sit closer to Hamilton, more distant from the speaker, with the lesser ranking taking seats toward the front. The lowly, as measured by the occasion, would migrate to the seats behind those at the table and arranged around the periphery of the room.

In the course of a week executives and middle managers might attend several meetings in this hallowed territory, each time sitting at a different station as the circumstances demanded. Being aware of this protocol, Decker was able to determine by observation the relative positions of those present, which, he assumed, included most if not all the senior staff. As the honored guest, he was offered a seat at the table at the immediate right of the throne.

"First, let me take a moment to welcome our distinguished visitor back to DSD after all these years," Hamilton began. "How long has it been, Jack, ten, eleven years? Time really flies when you're havin' fun, as they say. We're really pleased to have one of the top executives from Delaware Microwave visit us today to be brought up to speed on the SEANET program. Arnie Tell told us to give you the full dog and pony show, Jack, so we've gathered the top Division people and the key SEANET program team members to participate as we move through the day. Most of them are here for this brief introduction, then we'll split up and you'll see many of us again from time to time during the day as you make the rounds.

"I'd like to begin by introducing a few people you may not know, Jack." Hamilton continued by introducing about half the twenty or so people in the room, most with Vice President, Director, or Manager titles. Among those he introduced was Kenneth Martin, Director of the SEANET program, who would be, as Hamilton put it, "your baby-sitter during the tours today, Jack."

Decker was more than a little shocked. He had begun to suspect that these people, including Hamilton, did not know why he was here. Now he was certain. The introduction of the man he might potentially replace as his principal escort and companion for the day was pure Arnie Tell. Tell had not informed them, because if he had done so, the SEANET briefing might have been biased and guarded. He wasn't sure what purpose Tell had ascribed to his visit, but they were treating it quite seriously and pulling out all the stops. He decided to give no specific reason for being there either, because to do so might conflict with whatever Tell had told them. He was, at first, upset that Tell had not let him in on the strategy, but then he realized that it was probably some kind of a test for himself as well. Tell would, no doubt, get feedback from trusted agents on how he, Decker, had handled the situation.

Hamilton continued with a formal briefing utilizing an overhead projector with color transparencies to support his remarks. Viewgraphs, as these transparent charts were called in the business, were the standard visual presentation medium in the aerospace industry, inexpensive and easy to prepare with computer programs such as Power Point. Those used by Hamilton were, however, obviously done by the professional art staff of DSD.

First he presented the agenda for the day, which included this formal briefing employing several presenters, then would follow an informal meeting until noon with the SEANET Program Director, Ken Martin, and the SEANET Technical Director, Milt Karinski. Hamilton and a few members of his staff would join them for lunch in the Penthouse, and the afternoon would be spent touring the facilities and talking to people working on the various components of the SEANET system so that Decker could, as Hamilton put it, "... touch the hardware." At the end of the day Decker and Hamilton would "... spend a few quiet minutes together." This was standard practice for visits of this sort, the VIP and the top local executive exchanging views privately, giving each an opportunity to alert the other, as a professional courtesy, to any potential controversy which might arise from the visit.

The final part of Hamilton's briefing was the usual company organizational overview, showing how the division was organized and how SEANET fit into the big picture, ending with the SEANET project organization chart and the introduction of those present as their functions were addressed.

The next briefing contained a few surprises, as Harold Daley, the division's comptroller, gave a revealing picture of DSD's financial status and its dependency on SEANET's success, since SEANET represented almost half the division's present and future business. At the start of Daley's presentation, many in the audience departed quietly, by prearrangement Decker understood, because this kind of information was always closely held by the division's executives and financial department. The surprise was that it was being divulged to him, an outsider. What the hell had Arnie told them, he wondered. At any rate, Daley's message was clear: a major cost overrun on SEANET would wipe out half the division's profit for several years, and so far, things weren't going well. Decker detected a note of apprehension in the comptroller's voice as he reflected on the seriousness of the situation.

Ken Martin gave the SEANET Program briefing. He was an accomplished and effective speaker, demanded of a person in his position. He described the technical requirements of the program, much of which was classified secret, and he presented the master program schedule emphasizing the milestones projected over the next year. SEANET was an extremely ambitious undertaking involving a very large number of companies and government agencies, and it was no wonder that this first phase would cost a billion dollars and take years to accomplish. Martin also discussed the present status of the program and how DSD stood versus the planned cost and schedule objectives. Since he had, obviously, been instructed to hold nothing back, and despite practiced good cheer and an air of confidence, he was not successful in convincing Decker that the difficulties being encountered were under control.

Decker had been in Martin's position before and knew what to look for, but rather than ask incisive questions in front of the present audience of Martin's superiors, he decided to hold them for the private session to follow. Being late, being over budget, and encountering unexpected technical problems were not unusual occurrences in major aerospace development programs. The trick was to keep the three in proper balance so that the overall result at the end of the project came as close as possible to the original cost, schedule, and technical goals.

Decker believed that the taxpaying public, incited by politicians and the press, tended to overlook the enormous difficulty of developing new weapon systems and had come to believe that most high technology defense programs were either unnecessary, ill conceived, badly managed, too expensive, or all four. Some were. For the most part, Decker believed, those in his industry were conscientious, hard working, creative people trying to do a good job. As individuals they were not motivated by corporate greed, never inattentive to the public trust, nor dedicated to a perpetuation of international unrest and military conflict. Because of the high risk projects that had to be pursued to maintain technical dominance, however, there were frequent enough failures to satisfy the public's appetite for scandal, plenty of bureaucratic leaks to fuel the thirst of the press for sensationalism, and sufficient irresponsible politicians to join in the chorus of public outcry against fraud and waste, most of it unsubstantiated. Not to say that there were not incidents worthy of this treatment. On the other hand, to expect that projects like SEANET would produce exactly the anticipated results was not realistic. "We always set out to do more than we can realistically accomplish," Arnold Tell's father, Robert, had once told him. "If we didn't, we could never lead the pack."

Some were better at orchestrating and conducting these high-risk endeavors than others, as Arnold Tell realized, and that was why Jack Decker was here. Decker also knew he was one of the best, self aware of the rare blend of qualities that made him ideal for these jobs: intelligence, motivation, realism, organizational skill, political instinct, confidence, and a style of leadership based on communication and trust. Since teamwork was his fundamental strategy, his interpersonal managerial skills were his strongest asset, an ability to demand and receive the very best that each team member had to offer. He imagined that Ken Martin must have a large measure of these qualities also; otherwise he would never have been trusted with such an important project. Nevertheless, experience had shown that sometimes change was required just for the sake of change. SEANET might need only a fresh start in a new direction, a reassessment of objectives and a new plan for achieving them. Before this day was over he would know.

Milt Karinski's briefing was entirely technical, and in it he described the fundamental concept that had made SEANET possible. It was at once simple and elegant. There is a limit to the amount of information that can be communicated through wire or cable or space, because each independent signal has to be assigned to a channel wide enough to contain its information. Just as commercial radio and television stations are assigned to specific frequency channels, military communications channels are reserved for various kinds of transmissions. The amount of space occupied in the frequency spectrum by a given signal, or the width of its required channel, depends on its complexity -- simple Morse code signals requiring very little channel width, complex television signals consuming a hundred thousand times more. All of the Navy's required transmissions simply would not fit simultaneously into the limited amount of communications space, or bandwidth, available. What Karinski and his colleagues had developed was a method for compressing information and time-sharing the communications spectrum. Since most signals were not continuous and had unused spaces between bursts of true information, it was possible to pack other information into these unused time slots, however brief, to increase the total amount of information transmitted in a given bandwidth of frequency space.

The Karinski techniques required each signal to be analyzed by very high-speed computers, restructured and stored in such a way that it could be dispatched in small packets to fit into available time and frequency slots for transmission according to its priority. Under the most demanding traffic load there could be brief delays, a second or two perhaps, as computer controlled traffic cops selected and moved information over the Navy's communications alleys, streets, and highways. The practical bottom line result of Karinski's concept was a ten to one increase over present systems in peak traffic handling capability for a given amount of communications bandwidth.

There was an important side benefit to this method of spectral utilization. Since many signals were mixed together depending on their content in what appeared to be a random manner, they had to be unmixed in an exactly reverse process to derive the original information. This demanded special coding and decoding equipment that added a considerable measure of security to the transmissions. Classified signals could be encrypted as before, but even unencrypted signals were quite secure from all but the most sophisticated electronic eavesdroppers.

Karinski's research team had developed and proven the practical application of these concepts using Tellonics' own re-sources, and patents were held by the corporation for key ingredi-ents of the system. The relatively simple architecture of the SEANET system of communications satellites and compatible ground equipment, as complex as it was, depended entirely on the efficient use of spectrum made possible by the Tellonics research; otherwise, the number and complexity of satellites would have made the system unaffordable if not impractical altogether.

An experienced communications engineer, Jack Decker easily comprehended these basic concepts. As Karinski moved into a more specific but simplified discussion of the traffic management computer programs and coding-decoding equipment, Decker was able to follow the overview without difficulty. It was an impressive description of an important technological breakthrough of immense importance to the national defense. The system, when completed, would give the United States Navy an unprecedented advantage, so it was not surprising to learn that certain details of the system were classified Secret and some of the specifics, particularly the coding and decoding rules, Top Secret. The Top Secret matters were not discussed, since Decker's own clearance was limitd currently to Secret as documented in the classified visit request telefaxed to Tellonics prior to his departure from Philadel-phia.

Decker was very impressed with Karinski's work, and he said so. He remarked that he was glad "our side" had this technology, and that the Navy and DSD research team "deserve the country's applause, which, of course, they will never get." The acknowledgment of good work was as automatic to Decker as was breathing, not forced nor artificial, but practiced and comfortable. It was a manifestation of the personal management style that had served him so well over the years. Karinski sensed the sincerity of the compliment and was pleased, offering to pass it along to the other members of the research team.

It was almost exactly ten o'clock when Larry Hamilton returned to the podium to announce, "Well, Jack, that concludes this morning's formal briefings. I hope you've been able to get the flavor of what we're doing on SEANET; but, knowing you, Jack, I'm sure you've got a million questions. So now we're going to turn you over to Ken and Milt for the rest of the morning, and we'll see you again at lunch." Hamilton was right about Decker's accumulated store of questions, remembering well his appetite for developing a clear understanding of anything in which he had an interest. Hamilton had been uncertain when Arnie Tell had instructed him to "bare your soul on DSD and SEANET," but now he believed this to be a signal of Jack Decker's return to Tellonics.

"Fine, Larry," Decker replied. "I'd like to express my appreci-ation to all of you. It's been very enlightening for me, and I'm looking forward to the rest of the day. See you at lunch." With that, everyone rose from the table, and after a few minutes of small talk with those remaining in the room, Martin, Karinski and Decker departed together for Martin's office.

The next two hours would be the most critical in Jack Decker's decision-making process. During that time he must satisfy himself that SEANET had a reasonable chance of success. To reach that conclusion he would have to determine the true nature and difficulty of the technical problems, because engineering design was always the pacing factor. No one could schedule technical progress with certainty; however, one could assess the relative risk of technical issues, ascertain the availability of resources to resolve them, and determine if there would be enough time and money to do so before corporate management or the customer panicked, as they now appeared about to do. In addition, he would have to decide, ultimately, what to do with Ken Martin and Milt Karinski if he took over, so he needed to get at least a first impression of their current value to SEANET. He had already decided that he wanted the job. SEANET was the most exciting challenge he had ever confronted, but he would not ruin his good reputation and career by accepting a job that could not be done.

"It seems to me," Decker opened after the three of them had settled down with cups of black coffee in Martin's large but austere office, "that there must be a few tough technical problems that are holding up progress. Could you guys give me just the highlights of what they are?"

"You've hit that nail on the head, Jack. If we could just get a couple of subcontractors up to speed and resolve our syncodec stability problem, we could really move out and catch up." Decker had met Martin only a few hours ago, but, as was the custom in the business, they were already on a first name basis. "Why don't you explain the syncodec problem, Milt, then I'll talk a little about the subs."

Milt Karinski explained that the essential hardware element of the system was the top-secret synchronized-coder-decoder, or syncodec, which was being designed by DSD's data communications section. Research models and engineering prototypes had been built to demonstrate SEANET's fundamental data handling technology, but the initial production units did not have sufficient stability to operate over long periods of time without deterioration. "The surface and airborne syncodec designs look very good, Jack, but the satellite units are a tough problem. Our datacom engineers are the best in the business, but they need more time to solve this one."

He went on to explain that the satellite syncodecs had to be extremely compact and light weight, while at the same time they had to have the same kind of performance that the larger surface and airborne units had. It was essential that the entire system operate in perfect synchronization, and that required electronic speed and stability never before achieved in production hardware. "There are some promising solutions available, but they're all too big and too heavy. I'm afraid we just can't put a railroad locomotive into synchronous orbit just yet," he laughed.

Decker was thoroughly familiar with the necessity for engineering compromise. Sometimes it was not possible to meet all requirements simultaneously with available technology. "Push it in here, and it pokes out there," Arnold Tell had complained in frustration years ago when they were working together on a difficult circuit design problem. Engineering designs were often a set of acceptable compromises that came as close as possible to meeting all of the interactive requirements simultaneously. This appeared to be a classic case: high performance had to be traded for volume and weight.

"How long do you think it will take, Milt?" Decker inquired.

"We're trying out two new fixes in the lab," Karinski replied, "and both of them look pretty good. I'd say we need another month of lab work, and then we have to get a couple of production units modified. We need about six weeks before we have a production satellite syncodec that we can count on. It could take longer, of course."

Three months, Decker thought to himself. He tended to multiply engineering time estimates by two, because good engineers were always optimistic. They had to be, or they would never undertake difficult challenges. It was up to the program manager to be realistic and set aside a little time and money for a rainy day, because, sooner or later, it would certainly rain. Four months behind already, another three months delay to deal with, and a production model satellite to launch in a little over two months. It was not a good picture, but maybe it could all be made to fit by rearrangement of priorities and schedules.

"Are you certain that the syncodec is the long pole in the tent, Milt?" Decker suspected there were other technical problems. There always were, but the lesser ones tended to be obscured by the larger ones. As soon as one big problem was solved, there was usually another, almost as bad, to take its place.

"Yes, I really think it is, Jack. There are quite a few trouble-some design issues remaining, but none of them has resisted solution like this syncodec problem. As I said, if we can get the syncodec under control, I'll feel pretty good about the hardware." Decker believed that Karinski was leveling with him, and he was satisfied with the answer. On the other hand, he realized that problems that an engineer might believe were under control could prove otherwise. Engineers didn't tell you about those, usually in the sincere belief that they could be contained without outside help. Decker pondered, not if, but how many bad actors would appear without warning to spoil the play. If he took the job, he would ferret these out and see to it that attention was directed to each one.

"How about the software?" was Decker's next, incisive question, the inevitable one Martin had been waiting for. By software, Decker was referring to computer programs, which had, over the years, proven to be the most unpredictably difficult part of large system design. System engineers, software engineers and computer programmers were almost always too optimistic, and experience had taught Decker to be especially conservative in using their schedule estimates.

"Running a bit late," was Martin's response. "If it weren't for the syncodec problem, I expect we would be worrying more about software. The fact is, we are concerned, especially about the satellite software. Some of the traffic dispatching is done right in the satellite computers, and that's a tough requirement because they're smaller and slower. So far there are no show stoppers that we know about, but we're still about three months behind where we would like to be."

"Who's doing the satellite programming?" Decker asked.

"Commware is," Martin replied. "And that brings us around to the next subject, subcontractors. Commware is our largest sub -- about two hundred million altogether -- and it's mostly satellite software. They're programming the satellite computer, and we've got the system integration. Commware is so key to our success that we're practically living with them."

Decker recognized Commware as the company of Mason Crenshaw, about whom Arnie Tell had been obviously more concerned than he had admitted. Decker concluded that Crenshaw's own company being three months behind schedule was probably the reason Crenshaw hadn't raised more hell about SEANET in the boardroom.

"How are they to work with?" Decker continued this line of questioning to see where it would lead.

"It depends. The working level is top notch. They're probably the best in the business, and I've got a lot of confidence in their ability to deliver, even if they're late. We can live with that up to a point. But their management is a bitch to work with. If you don't dot every 'i' and cross every 't', you can bet on some kind of contract claim for cost or schedule relief. We have to run a pretty tight ship with them, so I've got a full time team working their subcontract just to keep us out of legal trouble. We should probably be more like that with our customers."

Decker took this offhand remark by Martin as an indication that he was having some trouble controlling changes by the Navy.

Arnie Tell had alluded to this also in the previous day's conversation. "I take it the Navy has a pretty long wish list."

"Sometimes I wonder whether we're working on the same program. My copy of the contract has a very rigorous change control procedure, but I think that page got left out of their copy," Martin complained, wryly. "Naturally, we resist, but on a cost-plus contract you've pretty well got to do what the customer wants." This was an inadvertent admission by Martin that he had been caving in to Navy demands for changes without getting proper credit in terms of recognized cost and schedule impact to the contract. It was the same old phenomenon that had led to serious difficulty in many defense contracts over the years.

Decker realized that it was always hard to say no to a good customer. He also understood the attitude of many government employees, military and civilian, that as long as the contractor got reimbursed for all of his costs, he shouldn't complain about a little change here and there. They could understand that if the contract price were fixed, the contractor might complain legitimately about changes in scope since it was "his" money. But when it was "our" money, what difference should it make to the contractor? What was misunderstood by the inexperienced, and often disregarded by the experienced, was that it was not "their" money and that the supply was not unlimited. Ultimately, contracts in which demanding customers had too much uncontrolled influence were headed for trouble.

"How is the Navy taking the overrun?" Decker continued.

Martin was unprepared for the question. While there was a potential cost problem, it was not serious in his view. "Overrun?" was his rhetorical reply.

"Well, if you're behind schedule and the customer is not paying for changes, you've got to have an overrun," was Decker's elaboration.

"Of course," Martin conceded. "We're slightly overspent for work accomplished to date, but both we and the Navy are pretty well convinced that we'll end up quite close to target. Frankly, Jack, they haven't given us too bad a time about it." Martin's response did not track with Arnie Tell's complaints about Admiral Sullivan's frequent phone calls. Something was wrong. He decided to drop the matter for the moment, but he would certainly ask Arnie Tell about it later.

The conversation continued for another hour, with several more cups of black coffee and the inevitable trips to the men's room in consequence. Decker spent the time looking at detailed schedules and budgets and actual contract performance measured against them. He concluded that the program planning was basically sound and that it might be possible to recover some of the lost time and prevent further financial erosion if tight controls were instituted. That could be done, however, only if the syncodec problem was solved soon and if the satellite computer programming remained on course. It was not an irresolvable risk on the surface, but what lie beneath the surface could not be discerned in a day.

———————————————————————————————

Lunch with Larry Hamilton and his staff was leisurely and pleasant. The Penthouse Restaurant was on the top floor of the executive office building, and while not of gourmet quality, the menu certainly could be rated on the high end of institutional fare. Not limited to executives, absolutely anyone in the company was welcome here; but the usual customers were executives, managers, engineers, or others who worked in the building. There was a rather nice company cafeteria in the main engineering building that wamuch less expensive and more popular, but visitors were usually treated to lunch in the Penthouse.

Hamilton had a permanently reserved table for eight at which he ate lunch most days. Others were free to join him without invitation unless, like today, a visitor was being hosted. It was a useful tradition, because unguarded comments made by this employee or that who might sit at Hamilton's table could be of great value to Hamilton. He believed in getting around and exchanging ideas with employees at every level when he could find the time to do it. Open invitation lunch was his principal means of bypassing the filtering of ideas that was inevitable in the upward flow of corporate information. There were a few, of course, who took advantage of the opportunity too often, and others who sought out these occasions for airing their personal grievances, which Hamilton despised but tolerated. The pluses far outweighed the minuses in his view, since everyone regarded the table talk at Hamilton's lunches as being off the record, and he was careful to threat it that way.

Decker noticed a few diners in uniform, mostly junior or staff-level naval officers, scattered about the room of a dozen tables. Military and civilian government employee guests were required by regulation to pay for their own meals. Most did, a few seldom did. It was a petty sin tolerated without question since those involved on either side hardly regarded it as the serious offense it was made out to be by government auditors and politicians. After each periodic scandal involving fraud or waste in government contracting, there would be a public display of conscience among the recipients of minor contractor gratuities, but it was usually short lived.

Out of the hundreds of government employees with whom he had shared a meal or a cocktail, Decker had encountered no more than two or three "straight arrows" who insisted on always paying their own way without exception. On the other hand, he had many casual and some very close government employee friends with whom he had worked over the years, and they were just as likely to buy him a drink as he, them. He regarded informal socializing with customers as an important communications medium, not unlike Larry Hamilton's open invitation lunches. In twenty-five years in the business, he had not once encountered personally a true case of bribery or influence peddling, although he believed that such things most likely did happen. But one thing was certain: the practice of the contractor's picking up the tab for lunch or dinner, once overlooked as a trivial matter, was now regarded with suspicion by the public at large and labeled as bribery by congress and the press. As a result, in recent months there had been a definite tightening of enforcement of the administrative regulations governing this practice.

The talk at Hamilton's table today centered, naturally, on Jack Decker and the events of the past decade. Decker had a chance to observe the others.who had joined him, Hamilton, Martin and Karinski. They were DSD vice presidents or department directors: Betty Emery of Human Relations, Stanley Abrahms of Engineering, Frank Cavallero of Operations, and Harold Daley, the Comptroller, whose financial presentation Decker had heard that morning. No one inquired about the purpose of Decker's visit, and he did not volunteer it. SEANET was not discussed, in part because of its classification and the public environment.

They were a compatible group, friendly, open, and with a good sense of humor to the man (and woman, of course). There was laughter and friendly sparring about who was responsible for this or that problem of the day. It was a phenomenon common to successful, high technology companies in which the work to be done was so complex and diverse that teamwork was absolutely essential to success. At the same time, the technical problems and business frustrations were frequently overwhelming, and only a bit of laughter could get one through the day. Decker had never met a good executive without a sense of humor, although some of the best wits were guarded in public and displayed only privately in order to preserve the public image of the owner as a serious minded executive.

****

After lunch, Milt Karinski excused himself to attend an important meeting after expressing his pleasure in having the opportunity to see Jack Decker again in such good health after all these years. They had not been close friends or frequent business associates, but their occasional professional encounters had been quite satisfactory for both of them, and each respected the other for his particular capabilities.

Ken Martin conducted the tour of SEANET facilities person- ally. In the executive office building was a display of SEANET hardware models and a very impressive, graphic presentation of the future SEANET system to be deployed around the world and in space. Next they visited the Engineering building in which Jack Decker had spent most of his last days with Tell Electronics. There, Decker was shown laboratory models of various equip- ments being developed by Tellonics. There was quite a large room in which engineering drawings of the interconnections of the entire system were laid out in detail around three of its four walls.

The engineers he met were bright and enthusiastic, most in their late thirties or early forties, and each appeared to be profes- sionally competent. These were the much coveted middle managers and experienced engineers, the backbone of the company, old enough to know how the business really worked and how to get things done, but still young enough to be innovative and in touch with state of the art technology. Decker understood that the kids and the old timers would be kept out of sight today. In Martin's place he would do the same thing, because the younger engineers might say something not in line with "the company position," while the old timers might be either cynical or complacent. They would know about and understand through experience the real impact of problems, however, and Decker wished he could have an hour with some of the more senior engineers his own age to get a feel for their comfort level with SEANET.

The next part of the tour was in the manufacturing facilities in which much of the SEANET equipment was being fabricated, assembled, and tested. It was impressive. The facilities had been constructed in the mid-eighties for just this kind of work. They were spacious, clean, laid out for efficient operations, and had sufficient high tech glitter to attract the caliber of technicians and craftsmen needed to build this kind of hardware. None of it was new to Decker, having spent a lot of time in these buildings, but he was impressed with the fact that there was a much higher level of production activity than when he had last worked there. People were everywhere, the machines were busy, and there was an atmosphere of controlled urgency and heightened activity in the place. He was surprised, because he had not believed that the SEANET equipment designs were far enough along for such a large-scale release to production, and he was concerned. If immature designs were being produced, there would be significant rework required if problems were discovered after the fact. Rework was expensive and time consuming.

The most impressive part of the tour was the building housing the "clean rooms" in which the most delicate equipment, especially the communications satellites, was assembled and tested. Millions had been invested in these buildings and their sophisticated environmental control systems in order to hold air purity, temperature, pressure, and humidity to exacting tolerances. There was less activity there, as Martin explained, because only one satellite was being assembled at the time. The production activity elsewhere was related mostly to the surface and airborne equipment to be used for the ships, submarines, aircraft, and shore stations.

One accumulated an extensive catalogue of names and faces over the years without really getting to know their owners. Decker recognized and was recognized by many of them as he and Martin moved through the plant. He met some new people, spoke to a few old acquaintances, and waved to many others. In this kind of organization, it was not possible to know everyone well, but each person accumulated a following of sorts and tended to know who they were. Decker had been an extremely popular engineer, then manager, and finally executive with Tell Electronics, and he was pleased to see that so many of his former support group were still with Tellonics.

At four o'clock they arrived back at the "brass and glass" building, and Ken Martin said good-bye to Jack Decker just outside Larry Hamilton's offices. Decker thanked him with sincerity for devoting his day to the visit, realizing that he must be very busy just now, and he wished the very best of luck to Martin and to SEANET.

They had been able to appraise each other rather well during the day. Martin was impressed with Decker's knowledge of the business, his quick comprehension of technical matters, and his intelligent, incisive questioning which stopped just short of nosiness. Decker liked Martin's grasp of the details of the job and his professional manner, but was uncertain about his sensitivity to some of the larger issues. All things considered, they had hit it off rather well. It was not clear how Martin was regarded by the customer or by top management, but he intended to explore that during his debriefing with Larry Hamilton and dinner with Arnie Tell.

Hamilton's first question was predictable. "Well, Jack, what do you think of SEANET?" His second was not. "And what the hell are you doing here?"

To the former, Decker replied that he was "... extremely impressed with the whole operation." To the latter he replied, "Larry, I think you'd better get that from Arnie. I will tell you one thing, though, I'm not after your job." They both chuckled, and Hamilton was somewhat relieved, knowing that Decker was certainly qualified for it. "Let's just say I'm a big fan of DSD, and I'm trying to find out how you guys do it so well." Decker seldom gave a "no comment" type of answer; neither did he answer delicate questions directly without careful consideration. On those occasions he just ignored the question asked and provided an answer for one that wasn't.

When Hamilton asked for Decker's candid opinion of SEANET status, he decided to be forthright for two reasons. First, he regarded Hamilton as a friend, and he would be doing him a disservice not to be an honest critic. Second, he knew he could trust Hamilton with off the record comments.

"I don't think you're doing too badly, Larry, but I do think it's going to get a lot worse before it gets better. You are late, of course, and you're going to have some cost problems as a result. If I was told about everything, you can probably prevent any sort of technical disaster if you work the problems hard, but you'd better tighten up on technical changes. It's time to cut the customer's phone lines, shoot the engineers and nail the drafting room door shut. I also think you'd better get some outside help with that syncodec problem. Milt Karinski and his people are good, but they may have a 'not invented here' attitude about changing some of their pet designs. I'm also pretty uneasy about the software that Commware is developing for the satellites. I didn't get good vibes on that at all, so a good Spanish Inquisition type program review with them might be a good idea. One other thing: production is pushing design, and you might have a rework problem downstream. That's about it. I'm probably not telling you anything you haven't already thought about, but that's my nutshell opinion as a seagull." Decker used the standard Navy definition of a seagull, with which they were both familiar, as a visitor who flies over the deck briefly, but leaves behind an unpleasant mess for others to clean up after he leaves.

"How about Ken Martin," Hamilton asked.

"Seems like a very competent guy. I liked him. He's right on top of everything and knows what's going on. I couldn't really tell if he's a driver or a passenger, though. What do you think, Larry?" Decker often ended his answers with a question in keeping with his general style of doing more listening than talking.

"He's our best. I believe he's done about as well as could be expected under the circumstances. Nobody's perfect, of course -- not even you and I, Jack," Hamilton chided, "but he's brought us a long way, and the customer thinks he's fantastic."

Decker's suspicions were confirmed. He knew that truly effective project managers were, as a rule, not altogether popular with customers. Respected, perhaps, but never viewed as "fantastic" unless they were giving away the store. They frequently had to give their own management a bad time as well to get things done for their projects. He asked Hamilton, "How much time do you spend with him, Larry?"

"He's got things under such good control that I only need to see him once or twice a week. He presents very thorough program analysis data at our monthly reviews, so we only talk about the exceptional things that may come to either of us from time to time. Of course, I've got my spies," Hamilton laughed, "so I'll get the word if anything goes seriously wrong."

Decker was aware of Hamilton's management style. He, too, was a delegator, giving his subordinates wide latitude in doing their own jobs, but Hamilton would, as he, himself, phrased it, "scratch around the barnyard to see out how the chickens are doing." It was a very effective approach in a manufacturing company where small problems could cause major impacts. That style forced his subordinates to dig into details lest they be confronted by Hamilton with bad news they did not already possess. Hamilton's subordinates were, as Decker had put it, on top of things; but since Martin's program accounted for almost half the division's business, it was very possible that he and the staff were fully occupied with detail and overlooking large matters of greater import.

The remaining few minutes of their conversation were personal. Decker inquired about the other Hamilton family members. He and Martha had been good friends of Larry and Jane Hamilton, and their kids had grown up together before Philadelphia. Decker contributed the status of his own three to the exchange. All of their kids were in or through college with the exception of Hamilton's youngest, who had been born slightly after the Deckers moved to Philadelphia, now ten years old. Decker remembered that Jane had been pregnant at Marty's funeral. The conversation brought back many of the events that Jack Decker had left California to forget, and he was surprised to find that most of the pain of those memories had subsided at last. Time did, indeed, cure all wounds.

****

Cindy Robbins and the silver limousine were waiting when Jack Decker stepped through the lobby doors of the brass and glass building, accompanied there by Larry Hamilton to signify his personal interest in his old friend's visit. As they said their good-byes, Hamilton recognized Cindy and waved to her, commenting to Decker, "Now that's my kind of bus driver." Decker smiled agreement, and Cindy returned the greeting with a knowing salute as she opened the rear door for Decker, who walked briskly to the curb in two-dozen steps.

Cindy described her busy day as they threaded their way through the rush hour traffic. She had been called back to Burbank to drive Arnie Tell and some VIPs to a luncheon meeting after it had been determined that Decker would not be needing the car until now. Then she had dropped off some visitors at LAX on her way back to Buena Park. Decker said little about his own day except that it had been quite full and a little tiring. He was glad it was over and was anxious to hit the pool at the Sheraton before preparing for dinner with Arnie Tell. He inquired if Cindy knew what the arrangements were.

"I was asked to tell you that Mr. Tell will pick you up at seven in the Lobby," she replied. He imagined that her invariant reference to her boss as "Mr. Tell" was carefully phrased to avoid any possible inference of other than a professional relationship.

The trip took only forty minutes this time. Jack shook Cindy's hand as he thanked her for all her help since his arrival. He was not certain if he would be using the car again, but he hoped to see her again before he returned to Philadelphia. She knew he would. Her instructions were to give top priority to Jack Decker's needs, whatever they were, but she did not disclose that fact. "I'll probably see you sometime tomorrow, Mr. Decker," were her final words as he exited and she closed the limousine door at the Sheraton entrance.

****

Seven o'clock came much too soon. Decker had arrived at the pool at five thirty, and after a few refreshing laps, relaxed for half an hour in the waning but still warm sunshine of the hot, dry August day with a draft beer. After a quick, steaming hot shower, he was dressed and in the lobby again to meet Arnie Tell at a minute before seven.

Tell walked through the same lobby doors Decker had used less than two hours before preceded by a large, well-dressed man he had not seen before. The man was introduced to Decker as "Mike Harrison, Cindy Robbins's nighttime replacement. Frankly, I'd rather have it the other way 'round," he joked as the three exchanged handshakes and pleasantries.

Arnold Tell treated everyone pretty much the same. It didn't matter if you were a senator or a gardener, conversations were conducted on a first name basis, and attention was paid to personal considerations. Tell didn't have an enemy in the world, he often boasted, except for "the bastards who're out to steal my money or my job." As chief executive of a large and powerful corporation, he was aware of plenty of people in those categories. Tell believed he could handle the business and the politics on his own, but he might need a little help in the physical department. For that reason, he never appeared in public without an armed Mike Harrison or Cindy Robbins close at hand. As a final measure of insurance, the cars he used, including the stretch Cadillac limousine, were armored and equipped with sophisticated security devices.

Tell erupted with, "I wanna know if you're gonna take the fucking job before I waste a lot of money on your goddam dinner, Decker!" He forgot to say bloody again.

To which Decker replied, "Well, I'd like to talk it over with Mike, here, before I decide." They all had a good laugh. Tell's jesting, big-bully exterior was well matched by Decker's quick wit. Their repartee' had delighted casual listeners for twenty years as it did Harrison, now, who laughed loudest of all.

In a black Mercedes 480SL, Harrison drove them a short distance, by California standards, to Albion's in Studio City, a small, quiet, continental restaurant of excellent and well deserved reputation. Decker knew that Mike Harrison could be trusted absolutely and that nothing need be held back in his presence. Otherwise, Arnie would not have him along. Harrison had a bright sense of humor as well, complaining that Tell was "a first class pain in the ass to keep up with, but I do get to hit all the top restaurants at his expense."

Harrison entered alone, and after a ten second survey of the interior of the restaurant from the entrance, nodded for Tell and Decker to follow.

After cocktails were ordered, Tell asked, simply, "Well?"

"I have to admit I'm very interested, Arnie, but I've got a couple of problems with it," Decker replied.

"Such as?"

"I can't figure out why you're getting all that heat from Admiral Sullivan, but the people at DSD don't seem to think the Navy has much of a problem. What's going on, Arnie?" Decker demanded.

"There are two problems, Jack," Tell replied. "First of all, Rusty Sullivan is trying to establish himself as a major program manager, and he's scared to death of making a mistake. He's determined that he's not going to be hung with a seagoing C-5A fiasco, so he panics at the slightest variance that exceeds one of those goddam cost-schedule-reporting thresholds. He's not a bad guy. In fact, he's smarter than hell and a hard charger. He just needs a little education and some handholding. What we've got to do is convince the sonovabitch we're not trying to screw him and to channel all that gold-braided energy into some productive activity. We've got to convince the guy to do his own job and let us do ours."

"And the second?" Decker prompted.

"They're related. We don't know what the fuck we're doing, and the sonovabitch smells it." These were bitter words coming from Arnie Tell, and Decker knew that he despised saying them. Because of his personal involvement in and commitment to the program, it was difficult for him to admit that SEANET was not going well.

"Larry Hamilton is the greatest plant manager in the world," he continued. "He can build anything, but he wouldn't recognize a bloody computer program if it bit him in the ass. He's just not a systems guy. And Ken Martin is the company's foremost paper pusher. He's got everything so organized that absolutely everything is accounted for. He can tell you to the minute how late any job is, and he can tell you to the penny how much cost growth we have on any piece of hardware. But there are four small items that he can't seem to account for: we're out of time, we're spending money like water, the customer's running amuck, and the fucking thing doesn't work. Aside from that, everything's terrific."

Arnie Tell's colorfully worded assessment matched Jack Decker's own in every respect. He required of CEO Arnie Tell this forthright acknowledgment of the true situation before announcing, "I'll take the job."

CHAPTER SIX
THE ADMIRAL

It had been just over two weeks since Jack Decker gave his answer to Arnie Tell. Immediately upon returning to Philadelphia, he had presented his resignation to Sam Edwards, president of Delaware Microwave, who accepted it "...with deep regret but complete understanding." After all, Edwards reasoned, Tellonics was one of his biggest customers, and it would be to his advantage to have his good friend, Jack Decker, in a position of authority there.

Decker knew that he was leaving the plant in good condition and that his departure with only two weeks notice would not be a hardship for Edwards. Although Edwards hated to loose Decker, a proven, steady hand at the wheel, it would not be difficult to replace him, since Decker had carefully prepared his subordinate, Ed Miller, to take over. He had developed the habit over the years of training one or more of his subordinates to replace him so that, in case a better opportunity should come his way, the lack of a replacement would not stand in the way. All that remained was for Edwards to agree to Miller's promotion, and he would very likely do so after going through the motions, as the Human Relations and Legal departments would demand, of interviewing other company prospects and, possibly, conducting a brief outside search for candidates. The sensitivity of corporate management to equal opportunity and affirmative action was, if not enthusiastically embraced, certainly institutionalized. It did not hurt that Miller was Black.

It had been a hectic two weeks. Decker had decided to leave the Bryn Mawr house intact, at least for a time, since the kids were all nearby and could keep an eye on it and possibly even use it. Depending on how things worked out downstream, he might sell the house later, and Tellonics would make him whole for any ex-penses or losses he might incur. To minimize commuting time, he would rent a small apartment or townhouse near the plant for a while, and consider a more permanent arrangement later. Orange County was awash with modern, luxurious apartments and condominiums, so it would not be difficult to find a suitable place to set up camp. That decided, there still remained a host of details, large and small, that required attention in effecting a move across country. Without neglecting the change of command at the factory, he had been absorbed with them.

His first official act as Vice President, Tellonics Corporation, and Director, SEANET Program, would not take place in Buena Park but in Washington, D.C. Arnie Tell had arranged to meet him in Washington, and together they would confer with Rear Admiral Russell "Rusty" Sullivan, providing an opportunity to introduce Decker to the Admiral as DSD's new SEANET Program Director so that he could, as Tell had phrased it, "... get the sonovabitch off my back."

Decker took the six thirty commuter to Ronald Reagan International Airport and then a taxi to nearby Crystal City, a complex of office buildings, high rise luxury apartments, hotels, and underground shopping malls in Arlington, Virginia, heavily populated with Navy agencies and their contractors. He met Arnie Tell at eight sharp, right on schedule, in the restaurant at the Hilton. Mike Harrison was there also, of course, and Decker mused that he would have preferred to see Cindy Robbins as bodyguard du jour even though Mike was a very likable guy.

As he and Harrison finished their breakfasts -- Decker had black coffee only -- Tell took the opportunity to bring Decker up to date on SEANET status which, in fact, had changed very little since their last meeting in California. They also reviewed their strategy for the meeting with the Admiral. Tell would do most of the talking at first, since he knew Sullivan personally and had communicated with him regularly of late. He wanted to make the point that Jack Decker was a very senior executive and a man he could deal with directly at his own level. Of course, he, Tell, would always be there if his direct attention were ever required, but hopefully, it would not be. Then the lead would switch to Decker. They would attempt to initiate a direct communication channel between Decker and Sullivan so that the two of them could run SEANET as a team effort with Sullivan taking the lead on the Government side, Decker on the industrial. Decker proposed, and Tell agreed, that they would, as a final item, ask the Admiral to be very candid and disclose his personal concerns so that Decker could go to work on the top problems as perceived by the Navy. Beyond contractual requirements, customer expectations had to be taken into account even if not specifically stated in the contract.

They also speculated on what the Navy wanted from this meeting. Tell had told the Admiral only that he was going to be in Washington and needed to meet with him personally to discuss some urgent business concerning SEANET. He would be bringing along one of his senior vice presidents. Rusty Sullivan had replied that it was a good idea for them to meet in person since he had some matters to discuss with him privately "... but not on the phone." Tell had concluded only that the Admiral must feel the need for a personal "ass-chewing" session with Tellonics' top man in order to bring his prime contractor up to flank speed. Decker offered aloud that, if that were Sullivan's idea, he would only be chewing on scar tissue since he, Tell, had been in that situation many times before.

They departed the Hilton at quarter 'till nine for the half-block walk to the Navy's National Center complex of office buildings. Mike Harrison stood aside after leading the way into building NC2 and was suddenly out of sight. He had arranged with the Navy Security Officer to keep an eye on his boss during their time in the building while he waited in the lobby. After signing in, they approached the elevators and were greeted by a handsome Lieutenant Commander in his early thirties, resplendent in short-sleeved, open-collar Navy whites. "Admiral Sullivan sends his compliments, Sir," he addressed Arnie Tell as they approached the elevators. "I'm Joe Alvarez. You may recall we met in Buena Park last spring at a SEANET program review."

"Of course, Joe. Nice to see you again. And thanks for picking us up. We poor contractors are not accustomed to such royal treatment," Tell offered, evincing a warm smile from Alvarez. He did, indeed, recall the young officer, if not by name. Alvarez, Sullivan's aide de camp, had given one of the Navy's briefings at the program review during the opening session which Tell had attended as a courtesy to the Admiral, and he had been very impressed with Alvarez's knowledge and enthusiasm. It was also his belief that officers of middle rank, gold and silver leaf level, really ran the military services. They had been around long enough to know what was going on and how to get things done in the bureaucracy, but not so long as to become entirely political and out of touch with reality. Almost all of the important actions taken by senior officers were based on the initiatives and recommendations of these bright young men and women.

"I'd like you to meet Jack Decker, one of our senior vice presidents," Tell continued.

Alvarez and Decker exchanged a hearty handshake. "Good morning, Commander," was Decker's greeting. Although a Lieutenant Commander, the abbreviated "Commander" was a customary salutation.

"Good morning, sir," was the reply. Decker was accustomed to being called, sir, but always felt a little uncomfortable when it happened. He preferred to communicate with everyone as an equal whatever their relative station in life and was usually successful in doing so, but he also realized that things were different in the military where rank was everything. It was Decker's observation after years of working with military people that they always sized up civilians and their professional positions, mentally ascribing an equivalent military rank to them as a means of establishing relative worth. He imagined that, in this situation, Alvarez probably rated Tell as a three star Admiral and himself,

Decker, as a senior Navy Captain. He set a mental objective of convincing Alvarez that he should be regarded as Rear Admiral equivalent in order to facilitate the establishment of a one-on-one relationship with Sullivan.

They arrived at the Admiral's reception room a few minutes before nine and were greeted by Jane Wilson, the admiral's civilian secretary. "The Admiral is expected just any minute from the morning briefings, so please have a seat in his office and I'll bring in the coffee right away."

Tell, Decker and Alvarez took seats at a small conference table opposite the Admiral's massive desk across the very large room, preset with Navy wardroom china cups and saucers and stainless teaspoons. The decor was all Navy, the walls decked out with the Admiral's accumulation of memorabilia from twenty five years as a naval officer, much of it spent at sea as a naval aviator and including command of a nuclear aircraft carrier, the Navy's most coveted assignment.

As Jane Wilson brought in a steaming hot pot of freshly brewed coffee, Arnie Tell struck up a polite conversation with her. They had talked many times as she located Tell for the Admiral's calls and had formed a relationship of sorts while exchanging pleasantries. Tell knew --as Decker had first brought to his attention years before -- that the only way to an executive's office is through his secretary, so it is important to have them as friends, never enemies. Tell had no difficulty with this state of affairs, frankly preferring the company of most of the secretaries he knew rather than their bosses. In the present case, he found Jane Wilson physically attractive as well as a pleasant foil for his frequent jests.

"Well, well! So Mister Tell finally comes to Washington!" The tall, redheaded admiral was cheerful as he entered his office, marching directly to Arnie Tell with extended hand. He, too, wore the informal white uniform of the day, with a chest full of colorfully striped ribbons and epaulets heavy with gold braid. He was an impressive figure, dominating the room as he entered, and they all stood out of respect not just for the rank, but also for the man.

"Good Morning, Admiral," replied Tell, half an inch taller, taking his hand with a vigorous squeeze, then, aside to Alvarez in an audible whisper, "Am I supposed to salute or something?"

"You're damned right, you'd better salute! This isn't the damned Marine Corps you know," Sullivan retorted, obviously aware of Tell's Viet Nam service. "If I had my way, all the civilians -- especially the ones who get elected to congress -- would be saluting me." Everyone had a good laugh, and Decker concluded from this friendly exchange that Tell and Sullivan had a totally open channel of communications despite Tell's desire to diminish the traffic. It was a matter of priorities with Tell, not preferences. Decker suspected that the Admiral's reference to congress was useful information also.

"Admiral Sullivan, I'd like you to meet Jack Decker," was Tell's introduction. The two exchanged greetings and a hand-shake, and the Admiral motioned for them all to take seats at the conference table, where he, rejecting the teacups, filled an over-size, white mug with the hot, black liquid from Jane Wilson's coffee pot.

"Hope you don't mind if Jose joins us?" queried the Admiral, referring to Lieutenant Commander Alvarez.

"Of course not," Tell replied. "Commander Alvarez and I are old acquaintances from California. Delighted to have him with us." He had inquired about Alvarez at the program review and was in-formed that he was the Admiral's aide in more than just a ceremo-nial sense. He was a trusted advisor on technical matters, an electrical engineer specializing in naval communications, and an academy man. More importantly, he was not buried in the official chain of command and had direct access to Sullivan on a continu-ing basis. It was important to have Joe Alvarez's support, so the more he knew about the management of the SEANET program the better advice he would give Sullivan.

"Well, Admiral, since I'm the one who asked for the meeting," Tell began in a more serious vein, "I guess it's my nickel, so I'll get right to the point. You know how important SEANET is to the Navy, and I want you to know it's important to us too. I believe we have to give it our very best shot, so I'm going to do whatever I have to do to see it succeed. It's an enormously complex program, and by God, it ought to be run by the very best Program Manager we can get our hands on. In my opinion, that's this guy over here, Jack Decker. So with your concurrence, Admiral, I'm going to appoint him Tellonics' SEANET Program Director."

All present realized that the decision had already been made. On the other hand, if the Navy had some legitimate objection to the contractor's choice for manager of one of its most important programs, they ought to have an opportunity to say so, and this was the time to do it, face to face with the principals. Tell continued, "I realize you don't know Jack personally and probably don't know his background, but I've known Jack since 1970, and I swear he's the most competent program manager I've ever worked with -- and I've seen a lot of them in my day. What's more, I've gotta tell you we're damned lucky he's willing to take over. It's no piece of cake."

"Well, congratulations, Mr. Decker," Sullivan responded with a smile. "Those are strong words coming from a man who's pretty stingy with a complement -- at least, I've never heard him say anything nice about anyone before. So what do you think about SEANET, Jack?" Sullivan usually addressed officers as well as ci-vilians as "mister," so the audience noted his friendly, offhand gesture of using Decker's first name.

Decker knew that his opening remarks would be critical to establishing a good relationship with Sullivan, so he had not left them to chance. He had planned them thoroughly, prepared to make adjustments as the circumstances demanded. There were several points he wanted to make, and he began with a personal note.

"First, Admiral, let me say that it's an honor for me to have been selected by Arnie to manage a program as important as SEANET -- and I am more amazed than you that he said those nice things about me. I only wish I had them on tape so I could play them back next time I get into serious trouble." The Admiral responded positively to the light humor, as Decker had calculated he would, and he continued. "At the same time, I must say that it is an awesome responsibility, second only to your own." This was an intentional deference to the Admiral's authority, but clearly not submissive as Decker continued with a "take charge" posture. "I was given the opportunity to look into the status of the program, so I know that SEANET has some serious problems, but I'm convinced that they can be solved without much further deteri-oration and that we can get there from here. I base that conclusion on the amount of progress you guys have made so far. But there are some changes that need to be made, and I think you ought to know about them before I make them."

"I'd certainly like to know what you recommend," Sullivan interposed, announcing that he was treating Decker's ideas only as recommendations at this point.

"So the hell would I," Tell chimed in, prompting another laugh around the table.

"Well, first we've got to get some help with the satellite syncodec problem. In my opinion that's the long pole in the tent. I don't have all the details because my top-secret clearance hasn't been reinstated yet, but I do know when to ask for help. Our engi-neers are some of the best in the business, but I think they're fresh out of ideas and a little stale on this one, so I'm going to bring in some outside consultants, possibly including some of your people, Admiral, and initiate a parallel program to work this problem." There was no reply from the others, but Decker detected a slight, positive nod of the Admiral's head at the invitation for the Navy to get involved intimately in the solution of the contractor's technical

76

problem. It was most unusual for a contractor to make such a gesture, and they all knew it.

"As far as cost control is concerned, with the exception of the syncodec, I didn't find any other bad actors, but rather a lot of small items adding up to a lot of money in the aggregate, and I've pretty well concluded that the cost overrun is mostly related to the schedule slips. It appears to me that we may just be victims of optimism during the proposal and program planning phases. I need to spend some more time on this, but my general approach will be to offload resources presently being used on SEANET until we catch up a bit on some of the key technical is sues. I think we may have been premature in starting production on some of the hardware in an attempt to maintain an unrealistic schedule, so some of that will have to be put on hold until we have a higher level of confidence in the design. That will cause us a little pain internally," he noted, glancing at Arnie Tell, "and some political pain for the Navy, perhaps, but we have to take steps in that direction. At any rate, as soon as we get the hardware bugs ironed out, we can turn on production full bore again and catch up. In the long run, it won't take any longer, but it will cost one hell of a lot less if we have fewer people standing around the factory waiting for the engineers to solve the design problems."

Decker's remarks so far made good sense to everyone. His next were challenges to both Sullivan and Tell. "To do that, we're going to need some help from the Navy, Admiral, and from Corpo-rate, Arnie." Decker glanced at each of them as he mentioned their names.

"First, we've got to rebaseline the program. We need to replan the whole thing together based on where we are today and where we've got to go. We need a new plan and schedule, and I'm willing to accept some tough challenges as long as they're realistic. We also need a new spending profile that matches the schedule. Moving money around from one fiscal year to another is going to cause some real dissent in the bureaucracy, Admiral, so if you

77

agree to this recovery plan, you're going to have to institute some cultural reforms in your organization. Two things have to happen there: one, help us plan out the program realistically and re-schedule all the things directly controlled by the Navy to match our schedule, and two, no more changes."

He paused for a response but, hearing none, continued. "I'm very serious about that, Admiral Sullivan. I'm putting out the word in California. I must personally approve every change, and no change will be approved without contractual cost and schedule consideration as our contract demands. For your part, Admiral, I'd recommend you or a trusted subordinate personally approve all requests for change before they are transmitted to us through contractual channels, rejecting anything but the real showstoppers." The Admiral was inscrutable. Decker hoped that his silence and serious posture were indicative of his contemplating the merits of these reforms with an open mind.

"On our side, Arnie," addressing his remarks to Tell, "we're going to have to make Mister Hamilton and Mister Daley a little un-happy for a while," referring to DSD's General Manager and Con-troller. It may not have been clear to Sullivan and Alvarez, but the two civilians understood that offloading production people from SEANET to other programs while engineering caught up would have a cost impact not directly chargeable to the contract except for its effect on the division's overhead which was prorated across all contracts. In the extreme, it could result in a substantial, if temporary, layoff at the DSD factory and at various subcontractors. This painful part of his plan had not been cleared in advance with Tell, Decker relying on Arnie's word that he would not interfere if he took the job. Tell remained poker faced but would, undoubtedly, have something to say once the customer was out of earshot.

"One other thing. I'd like to have Ken Martin and Milt Karinski stay on the program in essentially their same capacities for the time being, if they're willing. I'll be looking at the rest of the program staff in more detail after I report to California. I do

78

anticipate some measure of reorganization to get a little leaner and meaner, however, and that should happen in about a month. I'll touch base with you personally, Admiral, before I take any actions affecting key players that your people have been working with."

The piece de resistance followed, carefully orchestrated to achieve Arnie Tell's goal of offloading the Admiral to Decker.

"Along those lines," Decker continued, "I think it's essential for you and me to talk every day, Admiral. There's a lot going on, and we need to move in lock step for the next few months. I'd suggest you have Jane place a call every morning according to your schedule, and I'll make myself available. We don't need a fixed agenda, just talk about the worry list du jour. Frankly, Admiral, I can't manage a program if I'm not absolutely coordinated with my customer counterpart. I won't bore you with details, but I won't hold back anything important. As soon as I know I have a problem that could affect you, I'll tell you about it. In return, I definitely want you to share your problems with me to the extent you can." These last remarks were made to convince the Admiral to accept the contractor's Program Director, rather than the corporation's CEO, as his equal and principal point of contact. It was, in effect, a demotion in which national interest would have to trump personal pride.

The three waited for a response from the Admiral, who remained in studied silence for at least a full minute, eye to eye with Decker. "Well, Mr. Decker, I'd say you have pretty well hit the nail on the head. Your observations and conclusions are not unlike my own. In fact, it was my intention during this meeting to call Mr. Tell aside privately and give him one hell of a tongue lashing for not taking aboard more decisive management. It appears to me that he's done that already, so I'll have to forego the pleasure of using my colorful naval vocabulary at his expense. I can only hope that another suitable occasion will arise in the future so my speech will not go to waste."

"Not at my expense, I hope," Decker replied. The four laughed aloud at this good-natured exchange, three of them relieved at the Admiral's apparent acceptance of both Jack Decker and his strategy. Jose Alvarez was doubly pleased. He had been prodding the Admiral to have the contractor take positive corrective actions along the lines Decker was now proposing, and he had liked Decker instinctively at the outset. Decker's speech had not only confirmed his initial assessment but had also mentally promoted Decker to equivalent Rear Admiral.

Alvarez spoke for the first time since the Admiral had entered the room, supporting his superior by taking the initiative to propose a course of action. "I'd like to suggest that we send a small team out to DSD to work out the details of a new schedule and spending profile with you, Mr. Decker."

"Good idea, Commander," replied Decker, now clearly in charge. "Give us a week or so to get our own act together, then we'll schedule a joint planning session to work out the details. We'll need some of the major subs there also," he noted, implying that subcontractors such as Commware must participate in the re-planning of the program.

"I believe that Mister Alvarez has just volunteered to lead our delegation, Gentlemen. Right Jose?" asked the Admiral.

"Aye, aye, sir," was the obedient reply.

"Well, I think I can spare you for a few days," added Sullivan. "A man needs to get away from his protector now and again. Aides are pretty much like wives, you know. They tend to get underfoot after too much time ashore and start giving orders which we are compelled to obey." Arnie Tell understood, having had the same feelings about his bodyguards, even Cindy.

"There are, of course, some major difficulties from the Navy's point of view," Sullivan volunteered without prompting. "We do have our critics, as you know, and my staff and I spend more time defending this program than working on it. I guess that's just the nature of things in Washington these days, but what a monumental pain in the ass. They put me in charge of a two-billion-dollar aircraft carrier with five thousand sailors aboard and standing orders to start a war if I needed to, while in this billet I require daily congressional approval for every goddamn thing I do or say. But, politics is politics, and I accept it, so we're going to have to bare our souls when we come up with this new plan." Then, in language characteristically nautical, "I predict a gale force storm, about sea state four, when we hit those waters, so we'd better do a first class job of charting our course. I'll start testing the waters now, but when can we go public?"

"About a month," Decker replied, "and I think we can help with the politics. Arnie and I have a few friends in Washington, too, and I think they'll support us. We'll follow your lead on that, Admiral, and check with you first before we make any contacts. The main thing we have going for us, of course, is that SEANET could save this country one day, so it's worth a moderate delay and a small cost overrun. We've got to minimize it though, and we can only do this once. But I do agree that it's better to tell the world, voluntarily, that we have a problem rather than to let them find out anyway and accuse us of hiding the facts."

"You make a good point," added Sullivan. "We can only do this once. Let's be damned sure we get everything out on the table and not just industry's problems -- we've got our own dirty skivies to wash out here. Mister Alvarez," as he called Alvarez when in a serious mood, "set up a special staff meeting, and include all the agencies, too. I want to personally kick off this exercise to be sure they know how important it is. I want to be certain that we have security on this, too, until we're ready to release. No leaks! That goes for you guys, too, Decker."

"Understood, Admiral," Decker indicated.

"Glad to have you aboard, Mister Decker," Sullivan said as he rose from his chair, indicating that the meeting was at an end. The others rose as he raised the white mug in a simulated toast. "Here's to a successful mission. And I will call you every day. You can count on it. Jane will track you down like a bloodhound if you try to avoid me, you know."

As the three started toward the door, Sullivan addressed Tell directly. "Oh, by the way, Arnie, there's one other thing if you could spare another minute before you leave." Decker and Alvarez understood that Sullivan wanted a private word with Tell, so they continued to the reception room while Tell stayed behind. Decker also knew the meaning of "oh, by the way." The OBTW, as he called it, signaled the Admiral's real message that would be delivered to Arnie Tell alone in confidence.

Sullivan pushed the door closed as the two remained standing. "I think you've made a good choice in Decker. I certainly hope so, because he's got to save both our asses. I'll help him all I can, but as far as I'm concerned, you guys have fucked this program up from the beginning, so maybe now you can get it right. I'll go along this one last time, but if we don't make it now, I'm not going to be the courageous Captain who remains aboard and goes down alone with the ship. You're going down with me, Arnie. I promise you."

"Clear enough, Admiral. Not to worry. If anybody can do it, Decker can. But he can't do it alone. We'll all break our butts at Tellonics, but you'd better keep the fucking Navy out of our way." Tell was almost as surprised as Sullivan that he had been so candid.

"You do your job, and we'll do ours. I'll get the sailors under control, and the feather merchants, too," the latter expression Sullivan's pet phrase for the civil servants in his organization, "but that goddam Senate Armed Services Committee is something else. Blaisedale is closing in, so we'd better stop helping him and get this mess straightened out."

His reference was to Blake Blaisedale, senior staffer to Senator Joseph Flannery of Massachusetts and chairman of the Senate Armed Services Committee. Flannery was outraged when Tellonics was awarded SEANET. He had been assured by his confidants in the Department of Defense that the lucrative contract would go to a team headed by a major electronics firm in his constituency based on its low bid. He was embarrassed greatly when, at the last minute, the Navy source selection board, courageously, recommended an award to Tellonics based on its superior technical approach despite an estimated five percent higher overall cost to the government. Rusty Sullivan, as Source Selection Authority, had made the final decision. In revenge, Flannery had given direct orders to Blaisedale to monitor the contract closely and set up a deliberate campaign to kill the SEANET program and reopen it to competition at the first sign of serious trouble. Outside Flannery's immediate circle, only Sullivan and Tell were aware of this vendetta, and Tell had decided not to tell Decker about it.

Announcement of the appointment of a new top manager for an existing organization -- especially when a respected and well-liked manager is being replaced -- must be handled in such a way as to minimize disruption of ongoing activity. While Arnie Tell's bullish exterior, carefully cultivated, belied it, he was, in fact, an executive who cared about people, particularly those who had worked for him long and faithfully. Even if Jack Decker had not indicated a strong desire to have Ken Martin stay on as his deputy, Martin would have been treated with dignity and respect. As he always did when confronted with difficult personnel changes, Tell developed a strategy for handling this situation and, as always, rehearsed it mentally over a period of days. In addition, while less sensitive than the Martin move, personal talks with Larry Hamilton and Milt Karinski would be required before a general announcement was made. As Tell had been quoted, "It's okay to screw people as long as you tell them in advance you're gonna do it."

Rather than call them individually -- and alarmingly -- to his Burbank office, Tell decided to make one of his periodic, formal visits to Larry Hamilton's Defense Systems Division. It was his practice to visit each of the Tellonics divisions three or four times a year for a formal review of its operations. Major programs such as SEANET were, of course, always agenda items, so it would be an opportune time to announce the new organization. Usually, a month's notice of these visits was given in order to give the division time to tie any loose ends together and prepare formal presentations of the status of its operations, but in this case Tell would be conducting his review in less than two weeks -- having called Hamilton the day after Jack Decker accepted the position of SEANET Program Director. Tell's plan was to talk individually with Hamilton, Martin, and Karinski immediately upon early arrival at DSD and prior to the day's formal presentations, having arranged informally for their being available at an early hour. He would then announce to the staff at large the appointment of Jack Decker during the formal SEANET program review later in the day. He would depart for Washington the next day for the meeting with Decker and Admiral Sullivan.

It had all gone rather well, Tell reflected. Larry Hamilton was not surprised, he had admitted, having "... pretty well put two and two together ..." after Decker's mysterious visit. He was also very pleased about the appointment for two reasons: firstly, he and Decker were old and close friends who worked well together; and, secondly, he was glad to have some of the weight of SEANET lifted from his shoulders. Under the new organization, he and Decker would both report directly to Tell, Decker taking charge of the total SEANET program and Hamilton responsible only for the equipment which Defense Systems Division was building for the program -- plus, of course, all of his other responsibilities for other programs as DSD's General Manager. The financial aspects of SEANET, while managed by Decker, would remain under the DSD profit center umbrella for which Hamilton was still accountable. It was a divided responsibility that would have been uncomfortable for some executives, but Hamilton was content with the arrangement because of his personal confidence in Jack Decker.

Ken Martin was, at first, more than a little shocked and attempted, unsuccessfully, to disguise his disappointment. Tell made it quite clear, however, that Martin should not regard being Jack Decker's deputy as a demotion, and that he, Tell, would publicly announce that fact. Martin's position and duties would be largely unchanged, he understood, by Decker. Furthermore, by bringing in Decker as the intermediary between himself and Martin, he was, in effect, just replacing Larry Hamilton in that role. Thus, the SEANET program was being given corporate level rather than division level status. The program was so important to the corporation that it deserved its own corporate vice president, and Jack Decker was, as he must certainly agree, extremely well qualified for that role. Martin's own performance was not in question, and he would be recognized in time for his hard work.

The Tell-Martin interview ended with Martin's sincere offer to give Jack Decker every possible support in his new job. Martin, too, had formed a good opinion of Decker during their day together two weeks ago, and he, like Hamilton, was more than a little relieved to unload onto broader shoulders some of the burden he had been carrying.

Milt Karinski was flattered that Arnie Tell, as chief executive officer of multi-billion-dollar Tellonics, had taken the time and trouble to notify him personally and privately of the organizational change. Tell had anticipated that reaction precisely. Since Karinski's technical expertise was crucial to the program, the interview had been calculated to sustain Karinski's feeling of belonging, thus insuring that he and Decker would get off to a good start. Tell knew that, although they were not close friends, Karinski and Decker had worked well together in the past. As Tell had hoped, Karinski announced his pleasure at having Decker "... come back home."

The Principal Program Review, or PPR as it was called, required that the Program Manager and each of his principal team leaders present the status and outlook of their portions of the program. There was not much presented of which Arnie Tell was not already aware, but the occasion gave him an opportunity to critique the management of the program in as constructive a way as possible in a public forum. If a genuine "wood shed session," as Tell called it, be required, it would be conducted one-on-one with the recipient of Tell's displeasure. At the end of the formal SEANET presentation, Larry Hamilton announced that Arnie Tell had an important announcement to make.

"This is Tellonics' most important Government program," Tell began, "and we must work closely together to insure its success. For the most part, that is your responsibility, but corporate management has a responsibility, too. My job is to see to it that you have at your disposal all the tools you need to successfully execute the program, and one of those tools is the assignment of highly qualified personnel. I've done that. All of you were hand picked and personally approved by me, and you have not disappointed me. Under the circumstances, you have done about as well as could be reasonably expected. But as you have just confirmed, the program has some serious problems, and you need some help. To that end, I want to announce a slightly different organizational structure for SEANET." The audience became suddenly very alert. Organization changes were usually preceded by a few weeks of rumors and speculation, but Tell had caught them completely off guard. Until this morning, only he and Jack Decker were aware of it.

"First, let me make it clear that none of you is being replaced or demoted. For the immediate future, each of you will continue in his or her current role. What I am changing is the relationship between the SEANET program and corporate headquarters. Ken Martin will no longer report to Larry Hamilton as manager of a DSD program, but rather to a new corporate vice president who will report directly to me. You need to know that this reorganization has Larry Hamilton, Ken Martin, and Milt Karinski's enthusiastic support – and possibly their great relief." The audience chuckled lightly at this last.

"Having said that, and in view of recent events, it may come to you as no surprise that Jack Decker is coming home to Tellonics as Vice President and Director of the SEANET program. Many of you met Jack during his visit two weeks ago, and some of you worked with him when he was a Tellonics program manager on Northwind and other major programs. For those of you who don't know him, I'll just say that you are in for a great professional experience. I can think of no person in this business better qualified for this position. But you can judge that for yourself starting next Monday when he reports in. You also need to know that Jack and I will meet with Rear Admiral Sullivan tomorrow to inform him of the reorganization, and I anticipate that we will have his enthusiastic support. You have no better friend in Washington than our Admiral, and it is important that we live up to his expectations as well as mine. I firmly believe that with Jack Decker at the helm, we can do just that."

The meeting was over.

It was in this carefully tilled soil that Jack Decker would plant new seeds as Vice President, Tellonics Corporation, and Director, SEANET Program.

****

United 99 arrived at LAX on time again, and the ordeal of baggage claim and car rental was endured routinely. Decker could not help but reflect that there was no limousine and no Cindy Robbins to meet him this time, the later being the more disappointing. He was no longer a candidate for recruitment, of course, but an ordinary employee -- albeit a rather important one -- so he was entirely on his own this time. He had decided to spend the afternoon at the Tellonics' Burbank headquarters to take care of the various personnel check in procedures -- he was, of course, a corporate employee not a DSD employee -- spend a few minutes with Arnie Tell if he was available, and then proceed to the Marriott in Buena Park for his first official night back as a California resident. He would report in to DSD Monday morning.

Decker departed the Hertz rental car lot at half past noon, taking the San Diego freeway north and the Ventura east to the Lankershim exit toward Burbank. Knowing he would find a

McDonald's or Jack in the Box or Burger King along the way, he stopped for one of his favorite, if infrequent, luncheon pastimes, people watching over a burger and fries.

Decker believed that most people had a distorted view of corporate executives and would never expect to sit down beside one in a fast food restaurant. His own view of successful people at or above his own level of corporate management was that they were, in most respects, no different from himself. They had three distinguishing characteristics: they wanted to be in charge of whatever it was they were doing; they were willing to work a lot harder, smarter and longer for the privilege of doing so; and they were lucky. Of the three, lucky was probably the most important, because success was the coming together of preparation and opportunity -- the former, controllable, the latter not.

Eating a cheeseburger with strangers was relaxing to Decker, reinforcing his belief that he should not take himself too seriously because of the "lucky factor" in his career. Not to imply that Decker had a low opinion of himself, he knew exactly who he was and what he was capable of doing. But he also knew that truly extraordinary things could be accomplished only with the help of subordinates, so it was important not to feel superior just because he was lucky enough to be in charge. He was, in fact, contemptuous of those executives who were full of themselves.

Upon arrival at Tellonics corporate headquarters in Burbank, Decker was met by Tellonics' Vice President of Human Relations, Betty Emery, who accompanied him through the new employee sign in process. It took less than an hour since Decker received VIP treatment, which he greatly appreciated since he had much to do that day.

A call to Susan Anders to make an appointment with Arnie Tell was next on the agenda.

"I'm sorry, Jack, but Arnie's not back from Washington yet. In fact I haven't heard from him since his departure from the Hilton this morning." Anders sounded slightly alarmed. Regardless of his whereabouts and activities, Arnie Tell, when traveling, kept Susan Anders fully informed of his appointments and locations at which he could be reached if needed. Several in-flight phone conversations were the norm when returning from the East Coast, but none had been received. "The Sovereign departed Washington National at Noon, so I assume that Arnie was aboard and will arrive about three o'clock."

He was not aboard the Cessna Sovereign, one of four Tellonics corporate jets. Tell and Mike Harrison had stayed behind for a clandestine meeting in a nearby Alexandria hotel with the principals of Green-Masters Underwriters from New York. Arnie Tell was not one to leave his fate to destiny, especially when his management of Tellonics was being challenged. As a proactive defense, he was orchestrating a hostile takeover of Mason Crenshaw's Commware. If Tellonics owned Commware, Crenshaw would be replaced immediately as its CEO, and Crenshaw's clout on the Tellonics board would be considerably reduced. Not the slightest hint of that plan could be tolerated, and since Harrison believed that their movements might be tracked, the Sovereign jet had been sent home sans passengers for cover. It would change its flight plan en route and land at Pittsburgh, then return for them later that day after arrangements with Greene-Masters investment bankers were completed.

There was a secondary, but equally important, motivation for acquiring Commware. With Commware owned by Tellonics, the most troublesome subcontractor on SEANET could be managed directly by Jack Decker.

With the prospect of meeting with Arnie Tell indefinite, Decker set out for Orange County to do some apartment hunting.

CHAPTER EIGHT
DAY ONE

The first day on a new job is always uncertain, but Jack Decker was neither apprehensive nor unsure of his course of action. First, as a courtesy, he would meet briefly with Larry Hamilton. He anticipated that he would be escorted to an executive office, most likely a suite with a windowed corner office befitting a corporate vice president. He would accept it, but he would not use it except for special occasions. He would make arrangements for an alternate, modest workstation close to SEANET program activity. After Hamilton, he would meet with Ken Martin and Milt Karinski, and then they would be joined by the other members of the Program Management staff and the Department Team Leaders just to get acquainted. He would solicit informal comments from each of them in order to form an initial impression, but he would also schedule a formal standup presentation of program status by mid-week. The remainder of the day he would spend with Martin alone, familiarizing himself with the details of the contract, the technical specifications, and most importantly, the current program plan and progress against that plan. He anticipated that it would take three days to go through the entirety of the documentation at the work package level, calling in those responsible for critical work packages when needed to provide their assessments. This would be laborious, monotonous, tiring work, but Decker understood that a Program Manager's devils could only be found in those details.

He would need a first-rate administrative assistant who could be trusted to take care of the flow of non-critical work that would reach his desk so that he could concentrate on SEANET activities. He had several potential candidates in mind from his past association at DSD if they were available. First in line was his former secretary, Louise Davis. A call to Betty Emery at HR would be made to initiate that process. It was Tellonics policy to open new positions in the company to all qualified candidates who chose to apply, and Decker intended to interview all applicants before making a selection. Still, without prejudging the applicants, he hoped that Lou Davis would be among them.

"Welcome to DSD, Jack," was Larry Hamilton's greeting.

"Morning, Larry," Decker's response.

They shook hands, exchanged pleasantries, and started some shop talk centered around Decker's return to Tellonics and how things had changed in the eleven years he had been away. Foremost among the changes was the growth of Defense Systems Division's communications satellite business, an Arnold Tell strategic investment just getting underway when Decker had managed the Northwind program and had to rely on a subcontractor for the two satellites in that system. They had not talked for more than five minutes when Hamilton's phone rang. He answered it, listened, and handed the phone to Decker. "It's for you. Your Admiral."

"Jane Wilson here, Mr. Decker. I'll put the Admiral on." Wilson had started her search for Decker at Arnold Tell's office and was transferred to Hamilton's by Susan Anders. The standard protocol was for Wilson to get the recipient of the Admiral's call on the line first before putting Sullivan on the line. The exception of course, was anyone who outranked the Admiral. Contractors of any status were considered of lower rank and could be kept waiting.

"Good morning, Jack. Glad to see you hard at work on the Navy's business so bright and early. I'm just following orders, you see, the first of our daily phone calls." Admiral Sullivan appeared to be in good spirits this Monday morning.

"Thank you, sir. Jane tracked me down without too much difficulty I hope. I should have a regular battle station and telephone number by tomorrow, and I'll pass it along to her. How are things at the seat of the Government?" Decker was careful to exhibit some deference to the Admiral even though acknowledged as his contractor counterpart. The use of an occasional nautical term, like battle station, contributed to the posture.

"To tell you the truth, Jack, not much different than when we met last week. I'm pretty anxious to have our sailors and feather merchants go out to California to work with you on the replanning, though. When do you think you'll be ready?" the Admiral inquired.

"Let me suggest you have your people travel to California next Monday. By Tuesday we should have our first cut at a new schedule for their review and comment. I haven't really gotten into the details yet, but I do expect we'll need some major rework. I should be able to give you a sneak preview by the end of the week," Decker said.

"That's good," Sullivan responded. "I'll have Commander Alvarez contact you to make arrangements."

"I look forward to his call. And I also expect to have at least a rough estimate of any cost impact – and by the way, we'll be looking for cost reductions rather than cost growth beyond what has already occurred. No guarantees at this point, but that's the objective," Decker said.

"Well I sure hope so. The last thing I need on my plate is a trip to The Hill to ask for more funding. We're not too popular over there as you may have discovered." This was an understatement from Sullivan. He was certain that Senator Flannery was quietly waiting for disclosure of a major SEANET problem so that he could make trouble for the Navy and its selected prime contractor.

Decker suspected that the Navy, as was its usual prudent practice, had set aside a reserve to cover contractor overruns, certainly five percent, hopefully ten. He was determined to find a way to complete the contract within ten percent of the original target cost estimate, plus or minus negotiated changes that may have occurred since inception. At the moment, the projected overrun was fifteen percent, so some serious cost reduction effort would be required.

Decker continued, "And if I may be so bold, Admiral, we'll be looking for some possible reductions on your side of the fence also. In particular, I've been thinking about streamlining the test program -- with your concurrence, of course."

Sullivan was not offended. "If you can suggest ways for the Navy to be more cost effective, I want to hear about it, so don't be bashful.

"One last thing, Jack. I mentioned that we have to keep this replanning effort very quiet for the moment. Please caution your people, and inform only those with a need to know. We can't go public with this before laying some pipe here in Washington."

"I understand Admiral, and it will be done. Anything else today?" Decker asked.

"That's it for now, Jack. Thanks for the input, and I'll touch base with you tomorrow." Both Sullivan and Decker were pleased with the rapport that they had established so early in their association. Arnold Tell would be pleased also at the decreasing frequency of phone calls from Rusty Sullivan.

After the phone was handed back to Hamilton and hung up, he said, "That's a pretty big order, Jack. I wish you the best of luck, and we'll do everything we can to support you, but it will not be easy."

"Yes, I know. But we have no choice, and there will be some pain." Decker was dead serious about this, obviously his final statement on the matter, and Hamilton decided to leave it at that for the moment even though he might be affected personally.

Diane Foster, a planner from the Facilities Department, had been waiting outside Hamilton's office. Hamilton invited her in, introduced her to Decker, and asked her to show Decker to his new office and assist him with getting settled in. Foster led Decker a short distance from Hamilton's office to his own on the same floor. As expected, the walnut paneled corner office, nicely furnished with matching walnut furniture, was very much in the tradition of a top level corporate executive.

"This is really very nice, Diane. Did you make the arrangements and do the decorating?" Decker asked.

"Yes, sir, with the help of others, of course. We do hope you'll be comfortable here," Foster replied. "Is there anything else you'll be needing?"

"Well, let's see," Decker said as he looked around. His eye was drawn to the oversize walnut desk, atop which were an apparently new desktop personal computer and a combination printer/scanner/fax-machine. He noted the matching credenza that boasted a telephone instrument with myriad lights and buttons identifying several sequentially numbered phone lines plus intercommunication and speakerphone capability. "I assume the computer is connected to the corporate network."

"Yes, sir," Foster responded. "You have unlimited access to email to and from any employee at any Tellonics facility, and you have unfiltered access to the Internet and external email. There's a firewall, of course, to protect the corporate intranet from unauthorized outsiders. And by the way, your email account has already been set up – you're JackDecker@Tellonics.com. There's also a laptop that you can take along when traveling, and a cell phone with unlimited long distance and international access. You'll find them in the credenza. You'll have to input your passwords when you use the computers the first time. Both computers have a complete suite of office software -- we were told you prefer the Microsoft family."

"You seem to have thought of everything, Diane. It's really quite complete, and I do appreciate the trouble you've taken." Decker was accustomed to offering compliments for good work, but was never condescending nor insincere. Proportionate rebukes for mistakes were just as readily available in Decker's toolbox of interpersonal skills.

"There is another matter, though," Decker continued. "If it can be arranged, I need another modest work space near Ken Martin's general area. I may have to spend quite a bit of time there in the next few weeks, so I'll need another computer and a couple of phone lines that can be directly accessed from this office. My assistant will be in the outer office here most of the time, so she should be able reach me by intercom and transfer calls to me directly. And it should be more like a workstation than an executive office -- enclosed, of course, in case some privacy is needed, but nothing elaborate. Just plain vanilla will do nicely." Decker laid down the requirements for his real office.

"I think that can be arranged," said Foster. "I'll look around today for some suitable space and give you a call for your approval. After that, we'll need a couple of days to set it up. Will that be OK?"

"Perfect," said Decker. With that, Foster departed, very much impressed with the new corporate VP, not at all what she had expected.

Decker sat down in the comfortable leather swivel chair she had selected for him and addressed his new master on the massive desk – his personal computer. Decker mused that desktop computers didn't even exist when he first came to work for the old Tell Electronics, but in just twenty-five years they had become ubiquitous. Shipping clerks and top executives alike were now addicted to email, their corporate intranets, and the World Wide Web. Almost all communication and a very high percentage of routine business were now conducted while sitting in front of the screen. They were all slaves to this new technology, and Decker often wondered if they were more or less efficient than they had been previously in a paper-only world.

As Foster had promised, his new master required that he input various passwords which would be securely recorded on a remote server and used to regulate his access to company files or systems on the one hand, and to protect his private data on the other. He realized it was not very secure, but he always used the same password, ahtram – or ahtram1 if a number had to be included. It was his wife's name, Martha, spelled backwards.

He checked to see if by chance he had any email, expecting none, and was surprised to see about twenty messages. Two of them were announcements of meetings that he decided not to attend, there was a broadcast bulletin reminding everyone that the annual U.S. Savings Bond Drive was now in progress, and the remainder were personal notes welcoming him back to Tellonics. There was even one from Arnie Tell and another rather warm one from Arnie's secretary, Sue Anders. The greatest surprise was a brief note from radmrasullivan@navy.mil.gov. It stated simply, "Welcome aboard, Jack. Best regards, Rusty." Decker wondered if the Admiral's emails were all signed, Rusty. He took it as another sign that he had been accepted. It was a reminder not to screw up a good beginning.

Diane Foster had neatly arranged a few useful items on the walnut desk, among them a corporate phone book. Decker's first call was to Betty Emery to arrange for posting the administrative assistant position. "By the way, Betty, is Louise Davis still working here? She was my secretary before I left the company."

"I'm afraid I don't know her, personally, Jack, but I'll check. I assume you want her to apply if she can be made available. As you probably know, if the job would be a promotion for any of the applicants, the current boss is pretty well obligated to make her available with reasonable notice. In the meantime, would you like for me to arrange for someone to assist you temporarily until you make a decision." Emery was anxious to help Decker in any way she could without having to be asked.

"That would be very helpful, Betty. I won't be able to hide out for long, so the calls and paperwork will start rolling in within a day or two. Would you believe I even had email waiting for me this morning?" Decker was grateful for the offer of temporary help. He depended on his assistant to take care of routine matters and to use good judgment in passing things along for his personal attention, signature or action if required. Long ago he had abandoned the title, Secretary, in favor of Assistant, recognizing the important if unsung role of that position. He did not require a stenographer or typist. He generally typed his own documents on the computer (forty words per minute, 50 on a good day) thanks to having learned to type in high school, the only boy in the class. He needed good touch-typing speed in order to communicate with teletype in his amateur radio hobby, so it was worth the taunts of other boys who considered it a bit "sissy."

His routine was to handle each piece of paper only once if possible. He would have an "in" basket and an "out" basket, but no "pending" basket. There was a "read" basket, actually his briefcase, for special or lengthy items he would take home for information or study. His favorite basket was the "waste" basket, which he used frequently. He had found over the years that if he inadvertently threw away something that was important, another copy would appear in due course with a "second request" for action. Take care of it or get rid of it was Decker's practice. It was amazing how much time this acquired habit saved.

Next was a call to Ken Martin to set up the meetings with him, Milt Karinski, and the SEANET key personnel. For the later, a conference room would be required, and he asked Martin to make the arrangements. The private meetings with Martin and Karinski would be held in his new executive suite, that venue selected for a specific purpose. Decker believed it necessary to establish a clear line of authority, so it was necessary to have the first meeting with his new subordinates on his home turf. Subsequent day-to-day meetings could take place at his SEANET workstation, but the initial encounter had to be in VP country.

With no assistant on station outside Decker's office, a quiet knock on the open door at nine fifteen signaled Ken Martin's arrival. "Anybody home?"

"Hey, Ken! Come on in," Decker said with a broad smile as he turned away from the PC, arose from the leather chair, and walked toward Martin with hand extended. "Thanks for coming over."

"Pretty nice digs, Jack. I must say I'm impressed," Martin observed.

"Well, I'm impressed too. The facilities folks outdid themselves to provide this ostentatious little nest, but I don't expect to spend a lot of time here. I've asked for a small office down where the action is so that we can yell at each other more conveniently." Martin smiled at the thought of yelling at his new boss, but he took it as an invitation for open give and take whenever they might have differences – as they very likely would. It was precisely so intended.

They took seats on opposite sides of the big walnut desk, and Decker began. "Ken, I'd like to be sure we get off on the right foot and understand how we're going to work together, so let me give you my thoughts, and then I'd like to hear yours. Somebody has to be in charge, and I'm elected, but I'm not smart enough to direct this program without a hellavalot of help. Notice that I said, direct, not manage. I intend to set overall policy and give direction, but I'll trust you and others to manage it. Most of the time I will delegate responsibility and the authority to meet that responsibility, but I won't abandon ship. If there's a serious problem, I expect to be told about it, and I will get into it until it's solved. Can you live with that?" Decker was more interested in hearing Martin's thoughts than expounding his own.

"That won't be a problem for me, Jack," said Martin. "I'll take all the help I can get, and I could not have asked for more. Your reputation precedes you. You're a no-bullshit guy, and you won't get any from me and the others. Frankly, I look forward to working with you because you will have the clout to straighten out this program. There have been times when I just couldn't get the attention we needed from the front office, and that will be different now with you in charge. As for responsibility and authority, well, pile it on. I'm ready. And oh, one more thing. I will yell back when I think you're wrong."

Decker smiled, sensing correctly that Martin was sincere in his willingness to accept help from someone who had been there before. "Then we're definitely off on the right foot. Now where's that chief engineer hiding out?" he asked, referring to Karinski.

Martin placed a quick phone call to Milt Karinski, and expecting the call, Karinski appeared at the office door in minutes. They had worked together before, of course, so there was no formality between them. "Glad to have you back, Jack," was Karinski's sincere greeting.

"And I'm glad to be here, Milt." After a warm handshake, Decker motioned Karinski to an empty chair next to Martin. "You probably think we've been talking about you behind your back, Milt. And you're absolutely right," he joked.

Laughing, Karinski quipped, "Well, the rest of us have been talking about you, Jack. So I guess we're about even."
"Nothing too damning, I hope."

"Not at all," Karinski continued. "Mostly speculation on how things might change with a new Program Director coming on. The usual questions: what's he like? Is he as smart as they say? Will he reorganize the program staff? Stuff like that."

"And how did you answer?" Decker wanted to know.

"Well, let's see, " Karinski thought for a moment. "Cool, yes, and probably." They all had a good laugh.

Soon enough, they were into the status of the technical problems that had been brought to Decker's attention two weeks ago. Some progress had been made on repackaging the satellite syncodec, but the software problems were still a major concern. "We will solve these problems eventually," Karinski offered.

To which Martin replied, "Yes, but when, and at what cost?"

With no reply forthcoming from Karinski, Decker took the lead and stated, "Well, that's what we're all here to find out. I plan to make a bit of a speech when we get together with the staff," glancing at his watch and counting down to ten o'clock, "and this is what I'm going to say." Decker summarized briefly the agenda he had planned for the next two weeks. He would present it in some detail to the full staff, but he wanted to be sure that Martin and Karinski were not taken by surprise in front of their subordinates. "Any comments or suggestions?"

Martin and Karinski were, indeed, surprised, but better now than later. Martin's only comment was, "Those are big changes, Jack, but by God, taking the long term view is just what the program needs. Have you discussed all of this with Larry Hamilton?"

"Not yet," Decker replied, "but I think he suspects something along these lines. He'll come along."

Karinski added, "Some people will be upset, and I have a few questions myself, but you have my complete support. Engineers are optimists, as you know, but I really do think we can pull it off."

Martin, Karinski, and Decker entered the SEANET conference room, in that order, at exactly ten o'clock. The others were already present, chatting and having coffee, none wanting to be late for this first official encounter with the new VP. Martin introduced him to the dozen program team leaders, some of whom he had known previously, and others he had met during his visit two weeks ago. It was a very cordial welcome, and the brief conversations were a pleasant experience for all present.

When the introductions were completed, Martin requested, "Have a seat, please, everyone." When the audience was seated and quiet, he announced, "It's a great pleasure for me to introduce our new boss, Jack Decker. Many of you worked with Jack when he was here previously, and for those who have not yet worked with him, I believe you will be impressed. I know I have been. So without further ado, here's Jack."

Decker began, "Well, I appreciate those kind, if undeserved, words from Ken, but at least he didn't start with 'heeeeeer's Jackie!" The laughter was spontaneous and genuine. When it subsided, he continued, "And now I should say something like, 'but seriously,' and this is a serious business we're involved in. I'm confident that with all the talent in this room and elsewhere in these buildings, and with the Navy's strong support, we can complete this program essentially on schedule and within shouting distance of target cost. Right now I don't think that we're on track to do that, and I believe most of you would agree. The plan we are following now was a very good plan when the program started over two years ago, but it is no longer valid. It's time to put on the brakes, assess our actual status honestly, and replan the entire program in detail from this point forward to final acceptance by the Navy. And the Navy has agreed to help us do just that. Arnie Tell and I met with Admiral Sullivan in Washington on Thursday, and he is with us.

"I have some ideas, maybe a little screwy, maybe not, and so do Ken and Milt, on how we might proceed. But the really good ideas will come from the trenches, from the people charged with actually doing the work, and that means you, the members of this program team and your colleagues throughout the company." Decker knew from long experience that this had to be their plan, not his. They had to buy into it, or it would not work.

He continued, "Here's what we're going to do. I plan to spend the next couple of days locked in a room with Ken Martin learning this program like the back of my hand. We may call some of you in from time to time to discuss critical issues.

"Each of you will drop whatever you are doing, no matter how important you think it is, and spend the next two days making a fresh, honest, open-kimono assessment of where we stand today, and what we have remaining to do, in your area of responsibility. On Thursday, we will meet again in this room, and one by one you will stand up and tell the rest of us what you have found. I expect formal viewgraph presentations with facts and figures, risk assessments, cost estimates, all the other management tools available to you. Ken will work out the specific agenda with you at the end of the meeting.

"I expect there will be some issues on which we don't agree. We will debate those issues in an orderly way and make tentative decisions on how to proceed. When we are done – and warn the family you may be a little late on Thursday -- we will have the broad outlines of our new plan. Friday and Saturday – sorry about the Saturday – we will put flesh on those bones and give our plans to the program control folks who will input your new cost and schedule data into their computers and make the necessary adjustments or workarounds needed to get where we have to go. Program Control will have your preliminary cost and schedule status reports ready for you sometime Monday.

"The Navy, your counterparts, will be here on Tuesday, a week from tomorrow. We will sit down with them and get their comments and suggestions on our replanning, and we will take those comments and suggestions seriously so long as they are within the authorized scope of our contract. That will probably take two days, maybe three.

"I plan to meet first with Arnold Tell and then Admiral Sullivan at the end of next week to brief them on the new program plan and what we need from them to execute it. I expect the Navy team will have briefed the Admiral before I do, so it's important that we say the same thing. I have been assured of his cooperation so long as we stay within the confines of the contract and the technical specifications. Even those can be amended if it makes sense to do so.

"Let me conclude with a few suggestions -- well, maybe they are more than suggestions. Let's call them executive guidance – and a few observations from my own experience.

"Rule one. Don't leave anything out. Be sure everything is covered. The corollary is don't duplicate. Be sure we plan each thing only once.

"Rule two. Work the schedule to death. Make it as short as possible. The corollary here is don't be unrealistic. Allow some breathing room, but not much. Given the work to be done and the things to be delivered to the customer, cost is controlled primarily by controlling the schedule. The shorter the schedule, the less the cost.

"Rule three. Minimize the use of resources, especially labor. People should charge their time to this program only when they are performing value-added work on this program. Beyond a reasonable point, waiting around for parts to arrive, or watching someone else work, or standing by for engineering to redesign something -- that sort of non-value-added labor should be charged to overhead – not to your SEANET cost account. Overhead is Larry Hamilton's problem, not yours.

"Observation one. Avoid rework. When work is done incorrectly and has to be done over, it is pure waste. That does not apply just to the factory -- correcting a drawing or revising a report is rework. Releasing a design to production prematurely causes expensive rework. In other words, do it right the first time.

"Observation two, closely related to one. You cannot "inspect in" quality. Inspection, as far as I am concerned, is not value-added work. If things are done consistently right the first time, inspectors are not needed. My apologies to the Quality Control folks. We'll find value-added things for you to do instead.

"Observation three. If you get into trouble, ask for help. And keep on asking for help until you get it. This is the single, most-costly mistake that engineers – and, yes, program directors – can make. 'I can handle this myself, so I won't bother to report it,' may be the costliest statement most of us ever make. And, by the way, nothing personal, but software engineers are the worst in this category.

"Observation four, the last. We owe our customer everything he has contracted with us to do or to deliver – nothing less, but nothing more. We are not obligated to deliver more than has been contracted for, and if we do, we expend energy on unplanned activity, lengthening the schedule and adding to the customer's cost. I had this conversation with Admiral Sullivan, too. He agrees. If we add effort to the contract, we change the contract.

"Finally, I have an important constraint to place on you. It is essential that we keep the work we are about to do confidential. I don't mean confidential in the sense of a military security classification, I do mean closely held, contained within this management team, and not revealed to others without a need to know. You may have to bring in others to help, including our subcontractors, and if so they should be similarly cautioned. It is extremely important that our new plan, or even the fact that we are working on a new plan, not be made public until our management and the Navy has given it their final approval.

"Thank you all for your kind attention, and I'll be pleased to answer any questions you may have." Decker's speech was completed, and he was anxious to hear and observe the team's response.

There were no questions. As far as Decker was concerned, that was the worst possible response.

CHAPTER NINE
THE TEMP

Defense contracts come and go in the Southern California aerospace industry, and specialists migrate from job to job as the demand for their services moves from one company to the next. Temporary employees, "Temps" or "Job Shoppers," as they are respectfully called, work through agencies that specialize in various occupations – craftsmen, accountants, secretaries, technicians, housekeepers, engineers, administrators, lawyers, managers, even executives. They substitute for permanent employees during extended illnesses, maternity leave, sabbaticals, or unplanned departures, but most importantly in the defense business they populate companies whose permanent staff becomes overloaded when new contracts are booked. This is a good arrangement for employers since they assume no responsibility for employee benefits. They pay a fixed price to the agency for hours worked, and leave the burdens of payroll administration and employee benefits, if any, to the agency. The employer can end a service no longer required or reject an unsatisfactory performer without notice and without liability. On the other hand, if a temporary employee is found to be well qualified for a position, arrangements can be made, by mutual agreement of all parties, to make that person a permanent employee. The agency is paid a fee, the company has a new employee of demonstrated competence, and the employee has a permanent job with more security and better benefits.

Mechanical design draftsman Patrick Riley had been job shopping in Orange County for the last two years. Now forty, he was an experienced draftsman, but with a spotty employment record back in Massachusetts. Raised in a dysfunctional South Boston blue-collar household, young Patrick's family had provided little support for his education. Determined to become an engineer, he had found public financial support available, but his mediocre academic record and unremarkable SAT scores denied him entry to any of the engineering colleges or universities to which he had applied. He had settled for two years of junior college and then attended a design drafting school for another year, excelling in both. He found employment as a design draftsman without difficulty.

Riley was at a lower professional status than the engineers from who he received his assignments and direction, but he was convinced that he was just as intelligent and capable as they and a victim of circumstance. That may very well have been the case, but rather than continue his education toward an engineering degree, which his employers would have enthusiastically supported and financed, he became resentful and sullen. His work was of high quality, but his attitude resulted in such tension with supervisors and associates that he departed job after job within a year or two, each time seeking greener pastures.

Pat Riley had worked for five different companies in the twelve years after graduation from technical school, and the last was New England Electronics. When Tellonics won the SEANET competition, New England Electronics' plans were disrupted, and they were forced to lay off over one hundred people. Riley, with the least seniority in his department, was one of them. Discouraged, angry, with few prospects for employment in the area, and having burned a number of bridges behind him, he had expected to find more opportunity in California. Unfortunately, the defense business in California was at a low point, aggravated by the general migration of industry from California to more favorable business environments in nearby states. Nevertheless, experienced and holding a secret security clearance, Riley had been able to find regular work through Technical Specialist Company, a temporary employment agency in Anaheim.

Unable to form meaningful and enduring relationships with either men or women, Riley became a solitary outdoorsman as his principal diversion. Since coming to California he had revived that interest and taken up the related hobby of serious gun collecting. His collection of fine firearms included three pistols, his favorite a German Luger, a shotgun, two bolt-action hunting rifles, two semi-automatic hunting rifles, and one automatic rifle. He had purchased all of them legally and held valid licenses for those that required them. As was his nature, he had not associated with other hobbyists or their organizations, but he was widely read in their publications, many of which expressed an extreme political viewpoint. Indoctrinated with admiration for the Irish Republican Army in the Irish Catholic household of his youth, he found kinship with the Provos, the militant and more violent Provisional IRA. Over time he had become an anarchist with a genuine hatred of Government in general. With his career going nowhere, the military-industrial defense establishment, the source of his livelihood, became the focus of his discontent.

Six months before Jack Decker's arrival, Riley had been dispatched by Technical Specialist Company to Tellonics Defense Systems Division to work on the SEANET program. An expert at computer aided design, CAD, he was assigned to developing mechanical drawings for the SEANET ultra high frequency communications equipment. He was particularly adept at laying out multi-layer printed circuit boards with demanding tolerances. The quality and quantity of his work was so far above average, that he had been offered, and had accepted, a permanent position with DSD. He thought it ironic that the very company that had occasioned his bitter departure from New England was now his employer.

Riley decided to sabotage the SEANET program, and he believed that he was clever enough to get away with it.

CHAPTER TEN
BRAINSTORM

Thursday came quickly. Following the Monday morning meeting with the SEANET staff, Decker and Martin had spent that afternoon plus all of Tuesday and Wednesday reviewing the SEANET contract, its technical specifications, and most importantly, the breakdown of work to five or six levels of increasingly detailed subdivision. Viewed as a hierarchical pyramid, hundreds of individual work packages formed the base of the pyramid, which when summed up through levels of increasingly large combinations of related work, reached the apex of the pyramid at the top program element labeled, simply, SEANET. Each element at every level had associated with it a statement of work, a projected cost, and an estimated duration, and each was assigned to a responsible manager. The database containing those details was updated weekly by the responsible managers to show the actual cost and actual accomplishment to date such that one could see at a glance the current status of each element versus its plan, and each manager could see the various elements for which he was responsible. To process data of such complexity and magnitude, a very large, very fast, mainframe computer was dedicated to Program Control.

Decker did not need to examine every work element, because the Program Control computer with its specialized project management software was capable of identifying the elements that did not conform to plan, and Decker reviewed those elements without exception. The computer could also identify the critical path of the program schedule -- those elements of work that, when performed in the required sequence, made up the longest path through the interactive network of individual schedules. The critical path determined the final completion date of the total program. As expected, work elements defining software and syncodec packaging for the satellites that Milt Karinski had identified were right there on the critical path. There were additional paths through the maze of work package schedules with a total duration longer than could be tolerated. It was the nature of large programs that as soon as the problems of the critical path were resolved, another path would become the longest, thus critical, and would require similar attention. Eventually, with proper replanning, management attention, and application of resources, all paths through the network would conform to the required end date for the program. Decker examined the work to be done on every noncompliant schedule path with the same scrutiny as that on the current critical path.

Martin and Decker spent most of those two and a half twelve-hour days glued to the computer screen. Decker requested hard copy printouts of screens that he considered significant, and he annotated them with his observations as they exited Martin's desktop printer. At the end of the review, he had accumulated more than a hundred pages. The work had been intense, both mentally and physically tiring, but Decker believed he now had much of the information needed to address SEANET's most demanding problems. The solutions to those problems, by and large, would have to be supplied by the program team leaders and their colleagues, but at the total program level major course corrections would be required that only he, with Admiral Sullivan's concurrence, could order. That new top-level plan for SEANET was beginning to take shape in Decker's imagination, but he knew it would be a hard sell with the Navy.

Diane Foster called at mid-afternoon on Wednesday. She had found, and Decker approved, a suitable space for his "office away from the office," as he called it. She reported that it had been set up as he had requested and was ready for his use.

At the end of his SEANET review with Martin, by that time seven o'clock in the evening, he explored the alternate office with its efficient workstation arrangement. There was a personal computer identical to the one in his VP office, and a simpler telephone and intercom instrument. It was just as he had wanted it. He sent a thank you e-mail note to Diane Foster, and departed for home.

Until permanent quarters could be found, home was the Residence Inn in Anaheim Hills, a part of the Marriott Hotel Chain designed especially for business travelers staying more than a night or two and desiring the amenities of a home-like environment. Avoiding the complications of eating out, Decker placed his heavy briefcase on the coffee table in the living room area, turned on the television for background, placed a frozen pizza in the microwave in the galley-like kitchen, and changed into pajamas in the bedroom. He consumed the hot pizza with a cold Bass Ale while he watched the end of a "Law and Order" rerun that he had almost committed to memory. He wondered what the thoughts of his subordinates would be if they could observe their new VP and Program Director in his "executive suite." In the next few months they would come to know that he was not so different from themselves.

The next two hours were dedicated to a final look at the annotated pages from Ken Martin's printer in preparation for the meeting tomorrow morning. Decker retired at eleven thirty after watching the Channel Five news, slept without interruption, and was already awake at six-thirty Thursday morning when the wake-up call came. After energizing the coffee maker in the galley, pre-loaded the night before, Decker shaved, showered, and dressed. For breakfast he had a bowl of Grape Nuts with sliced banana and milk and two mugs of black coffee, the second accompanying a quick scan of the "USA Today" found just outside the entry door. Off to the office at seven thirty, he arrived at the Glass and Brass building at seven fifty to attend the eight o'clock meeting to end all meetings.

Decker had become a believer in the technique known as brainstorming. When working on a problem as an individual, one tended to settle in on a prefered solution based on knowledge and prior experience. Others not so intimately involved might see things differently and propose alternate solutions based on their different base of knowledge and experience. No limits on plausibility or practicality are imporsed, and even a solution originally viewed as rediculous might win the day.

"Good morning, everyone, " Decker greeted the staff as he took charge of the meeting personally. "Ken and I spent the last couple of days going over the program in some detail. I believe I now have a better understanding of where we stand today, and I've seen many of the problems you are facing. The purpose of this meeting, as I indicated on Monday, is to identify those problems and take at least a first cut at their solutions. There will be no blame or recrimination for things that may have gone wrong in the past, and certainly no scapegoats appointed. So I'm expecting candid disclosure from each of you, and then we'll work the problems together. At the end of the day, and taking all of the issues into account, we should be able to establish a new plan with high confidence."

"I know you put a lot of work into preparation for today on very short notice, but today, please skip over the easy parts, and get right to the heart of the matter. What's wrong, and how can we fix it? "

The first presenter was Milt Karinski as the principal system engineer responsible for the total system's performance. Karinski was followed in turn by presenters responsible for the subsystems of SEANET. The Navy's top-level specification defined the technical requirements for the total system, and it had been partitioned by Karinski's system engineers into subsystems. Individual specifications were written for each subsystem based on its required contributions to the total system. Those subsystems, in turn, were divided into components each with its own specification, and so on down the line to individual physical parts or units of software. If each part, unit, component, and subsystem performed to its own specification, then in theory the total system would come together and meet the total system specification.

As each presenter concluded his presentation of issues requiring resolution, Decker's strategy was to ask for suggestions from the other participants. While there was initially some reluctance to speak up, Decker's prodding eventually led to a free exchange of ideas. It was this multiplicity of ideas that come forward in a non-threatening environment that Decker hoped to find new approaches to old problems. Typical of the exchanges was that related to a computer speed problem presented by Karinski.

He began, "Jack, as you requested I'm going to spare you the usual review of the specification tree and take you directly to a couple of items that do not currently meet their technical requirements and that have a direct impact on total system performance." Karinski flipped through his first several viewgraphs without pause or comment and stopped at about the tenth, placing it on the projector. It was captioned, Satellite Software. "Here's the number one system problem," Karinski noted. His comments followed the order of the cryptic sentences on the viewgraph, each preceded by a symbol, or "bullet," to introduce the next thought in logical sequence.

"The satellite hardware is mostly transmitters and receivers, but the satellite computers are the traffic cops of the system. The software in those computers must identify brief packets of information received from any network source, associate it with that source and the intended recipient, and send it on its way to the proper destination. It's similar to the Internet, but very high speed and very high capacity. The satellite can receive thousands of tiny packets of information coming in simultaneously over multiple prearranged channels. For security, those channels are constantly being reassigned in a pseudo-random fashion but synchronized with the users of the network. The software that does that channel scheduling and message routing is quite large in scale, but it is pretty much straightforward bookkeeping. That part is working as intended.

"The hard part is the software that de-encrypts the packets, stores them, associates them with the other packets for that message, rearranges them in the correct order, re-encrypts them, and passes the new packets along to the bookkeeper for transmission. All of this has to be done without significant delay. That software basically works, but at this point it can't handle the specified maximum traffic load – only about sixty percent. And that's the problem.

"Several solutions have been suggested. One, keep the software pretty much as it is, but get a faster computer. Two, add one or more additional computers and divide the traffic among the computers operating in parallel. Three, rewrite the software in a different programming language that will run more efficiently in the computer. Number one is a good idea, but we're using the fastest military microprocessors currently available. A year from now a suitable microprocessor might become available, or maybe not. Number two could work possibly, but there's no more physical space available in the platform. I like number three. I'm convinced that if we rewrite the software in C++ rather than Ada, we can meet the speed requirement. The problem is it will take too long and cost too much, and we'd need a contract change to use C++. As you pointed out Monday, Jack, this would be rework – redoing expensive work already done and paid for."

The SEANET contract, like many of its predecessors, required that software be written using a programming language called "Ada," which was developed by the Department of Defense in the seventies for universal military use. It was named for Augusta Ada Byron, the daughter of English poet Lord Byron and a nineteenth century mathematical genius credited with the initial concept of computer programming. Ada's principal features are that it produces software that is reliable, stable, and readily modified without unintended consequences. On the other hand, "C" is a proprietary programming language developed for commercial applications, now in its third generation of evolution and called C++, pronounced "see-plus-plus." It had been demonstrated that software written in C++ runs much faster than Ada software of identical functionality. In addition, with disciplined control of the software development process, software written in C++ can also be sufficiently reliable and stable for military use. There are other programming languages with similar virtues. Consequently, although still widely used, few military contracts now require the use of Ada exclusively.

"In short," Karinski concluded this topic, "none of the proposed solutions is promising, so we're still looking for a solution that can be implemented without serious impact on cost or schedule."

Decker was quick to concur. "Don't worry about insulting Milt, and don't worry about sounding ridiculous. We won't laugh. Sometimes a wild brainstorm really works. Any suggestions?"

All present stared intently at Karinski's viewgraph projection on the white screen. Half a minute of seemingly interminable silence had passed, when Mary Stevens spoke up. "I wonder if it would be possible to develop an Ada to C++ translator." Mary Stevens was a middle manager in the Engineering Department responsible for development of software used in SEANET's surface ship equipment. She was also the Engineering Department's software development lead on the SEANET program team and would be one of today's presenters.

"In other words," Stevens explained, "since the Ada software already exists and works, but is too slow, could we develop a program that converts the Ada commands to produce equivalent C++ commands that do the same thing but faster?"

"That's an interesting concept," contributed Morton Drake, lead representative from the Satellite Transceiver Section of the Engineering Department, "but as I understand it, there are major differences in the way Ada and C++ handle data structures. How would you deal with that?"

"I realize this is an oversimplification, " Stevens responded. "There are lots of nooks and crannies that we'd have to look into – and data structures is certainly one of them. But it occurs to me that there is a finite number of differences, and so long as we identify them and take them all into account, we might be able to write such a translator."

Decker, welcoming the exchange, interjected, "I wonder if such a translator already exits somewhere. We may not be the only project that has run into this problem. Anybody know of such a thing?"

With no response forthcoming, Karinski said, "Well, we can certainly do a bit of research and find out. Meanwhile, I think your idea has some promise, Mary. My suggestion would be to have a small group look into it and come up with a recommendation. We'd need an Ada expert and a C++ expert – perhaps more than one each – and a software system engineer to insure we don't overlook anything. Let's at least check out the feasibility."

"So be it," said Decker. "Ken," addressing Martin, "get with the leads at the break and identify the best possible members for a study. I'll contact Engineering management and get them assigned to work on it right away. We need at least a recommendation no later than Monday before the Navy arrives.

"Anyone else have a solution for Doctor Karinski's dilemma?" At carefully selected times, Decker would attempt to

117

level the playing field so that all participants would feel of equal importance. In addressing Milt Karinski as "Doctor," but without emphasis, he conveyed the notion that even a Ph.D. could have a problem that might best be solved with the assistance of lesser mortals.

And so it went throughout the day. The brainstorming among this multi-disciplined cadre of engineers and scientists was producing the intended results: new ways to look at old problems. Frequently more than one idea needed further exploration. In the case of Karinski's software speed problem, Decker added his own suggestion. "I may be able to help a little. I've got some contacts at Zybyte. They're developing their next, faster processor. I'll try to find out if they have anything experimental that we might be able to use. I'll let you know what I find out."

As the presentations proceeded, Decker's apprehension at the end of the Monday meeting had been found to be unwarranted. Ken Martin's observation, when asked about the absence of questions, was that they were either so well instructed they needed no further direction, or that they were awe-struck and reluctant to speak up. Whichever might have been the case then, it did not matter now. It was apparent that the program team was competent, bright, connected, and enthusiastic. At the same time Decker realized that making an initial formal presentation to a new superior was a frightening experience for some, and at least worrisome for most. He did whatever he could to put the presenters at ease and to support them, but he did not refrain from questioning, offering constructive criticism and inviting that of others. The pattern had been set with the relaxed presentation of Karinski, an experienced presenter with a prior association with the new boss. That presentation was now viewed by those present as a very constructive event, and they all hoped that their own would be as successful.

Decker, Martin, and Karinski sat together in the DSD cafeteria for a quick lunch. It was a welcome break from the mentally intense demands of the morning, so they did not talk business. The conversation centered instead on Decker's relocation from Philadelphia. He told them that he had seen a few condos and apartments over the weekend, and had found a nicely furnished townhouse that he might lease for a year until more permanent arrangements could be made, "if I don't get fired beforehand," he quipped. They all doubted that would happen.

The afternoon session began on schedule at one thirty, the lunch break shortened to thirty minutes to make up lost time.

The next presentations were made by the SEANET team leaders responsible collectively for producing the satellites and the communications equipment for the various platforms. There were inflexible dates for delivery of equipment to the shipyards for installation in the surface ships and submarines, and similar demanding delivery dates for the aircraft and shore stations. The satellite deliveries were carefully scheduled in coordination with Vandenburg Air Force Base, from which they would be launched, and once established, they were considered to be cast in concrete. A common Manufacturing Resource Planning computer program, MRP, computed the required availability date of every part, the conduct of every manufacturing step, the performance of every inspection or test.

Robert Masters, responsible for both material purchasing and SEANET subcontracts began with a listing of the thirty-nine subcontractors in descending order of subcontract value. Subcontractors were, by definition, those suppliers with either very large-value orders from DSD or suppliers developing new products or computer programs that could not be ordered "off-the-shelf." Listed first was the subcontract to Commware for the satellite software with a value of eighty-seven million, five hundred twelve thousand dollars. The Commware subcontract, like the SEANET prime contract, was a cost reimbursement type because of its high risk. The remaining subcontracts, the next largest of which was just under fifteen million dollars, were a mixture of cost reimbursement and fixed price contracts. Of the smaller subcontracts, three had relatively minor schedule problems, easily accommodated, and two had material quality problems that were being actively worked by both subcontractor and DSD engineers. "The big ticket item," as Masters phrased it, "is Commware.

"Ordinarily we would have developed the satellite software in-house, but as Mary Stevens pointed out, we barely have enough engineers to develop the non-satellite software, so we subcontracted the satellite to Commware. They're a very capable company, and we've done business with them before. Their performance to date is not only off the mark on schedule, they also have a pretty serious cost overrun. We had set aside a reserve to cover a potential overrun, but I'm afraid they are already well beyond that – about twenty percent projected at this point. So we have, roughly, a twelve million dollar problem, and the Navy sees that at about fifteen million. Not a big percentage of our billion dollar prime contract, but enough to attract our attention and theirs."

"What's the cause of the overrun?" Decker asked.

"No one thing in particular," Masters said. "The overrun is spread across the board uniformly. It's such an unusual situation that we have had a number of cost reviews with them. They have not been very forthcoming, and we don't have a lot of leverage to get at the real facts. They just say, 'Well, it's taking longer and costing more than we thought it would.'"

Decker suspected what might have happened. The fact that the overrun was uniformly distributed across all the units of software suggested that Commware had made a bottom line management adjustment to their engineering cost estimate, intentionally reducing the estimated complexity for all software units to achieve the desired number. It was not clear why Commware might have submitted an intentionally low estimate, but Decker intended to share this insight with Arnie Tell.

Returning to Masters, Decker asked, "You mentioned, Bob, that they were not forthcoming. Is that correct?"

"Right." Masters said. "Getting information out of them is like herding cats. We've had to resort to formal written requests, and even those are not always answered to our satisfaction. And don't bother to ask for any sort of change without full documentation and cost adjustment. Those guys are playing serious hard ball."

Decker was reminded of his conversation with Admiral Sullivan on the subject of undocumented contract changes. The phrase that came to mind was: what's sauce for the goose is sauce for the gander.

Decker asked next, "Do we have any offsetting cost savings?"

"Not in the subcontracts," Masters said, "but we're doing well with our material bid-to buy -- but not quite enough to cover it." Masters referred to the difference between the estimated cost of purchased material versus the actual cost – the bid-to-buy savings. These were typically achieved through creative purchasing agreements and tough but fair negotiating with suppliers.

"OK. I get the picture." Decker directed, "Continue to work the cost problem and do the best you can, but one thing we must do, Bob. I really want you to hold their feet to the fire on schedule. Don't give them a single day. Those satellites must, repeat must, be ready in time for launch. You can't make software changes in deep space -- at least not major ones. And just to help make the point, I'd like you to schedule a formal program review with them in about two weeks. Our place or theirs, I will attend."

Master's next viewgraph identified the top twenty-five vendors other than subcontractors. The most expensive parts, costing sixty five thousand dollars each, was the laser gyros. Two were required to stabilize each of the three satellites, a total of six. The least expensive part on the top-twenty-five list was the surface acoustic wave filters used in the communications receivers. Three hundred fifty of those were needed at a cost of sixteen hundred fifty-four dollars each.

Masters paraphrased the famous quotation attributed to the late Senator Everett Dirksen, reflecting on the multi-million dollar wish lists of his colleagues. "As Senator Everett Dirksen might have observed, it all adds up -- a million here, a million there. Although I believe he said billion, not million. At any rate, the total value of purchase orders and subcontracts that we have issued to date is a little over three hundred million dollars. To that figure add another couple of million for purchased services, such as specialty machine shops, testing laboratories, and chemical treatment facilities beyond the in-house capability of DSD." Decker was not surprised by the numbers, about a quarter of the one point two billion dollar value of the SEANET prime contract. That seemed about right. Even so, these were extraordinary statistics and managing this magnitude of purchased material was no job for lightweights

Masters continued, "You will not be surprised to hear that we have a few problems. I'll skip over the easy ones as you requested – I think we can either improve or work around those, about a hundred fifty of them are concerns, and they are all listed in your review package if you need more information. The hard ones are those with no solution in sight that seriously threaten delivery dates. While I don't want to point fingers at the engineers – they've been damn good on this job – twenty-seven items are late purchases due to engineering changes. And there are five more due to unacceptable material requiring rework or rebuys. So there are thirty-two top priority problems that you need to help us worry about, Jack."

Decker liked Masters style. He was professional and very much on top of things but was willing to ask for help when he needed it. As Masters went through the list of thirty-two priority material problems, Decker knew he could help and mentally formed a plan for doing so. Very often, he had found, a call from one top executive to another, and especially from a valued customer, was all that was needed to get the necessary resources applied to solving a problem. He also knew that engineers frequently took the easy way out of a manufacturing problem by changing parts used in the design. This could have unintended consequences downstream, so Decker would take engineering to task on each of those changes to see if another solution were possible. Masters would get the help he needed.

Ralph Brown, Manager of SEANET Manufacturing, was the person most directly responsible for building reliable hardware suitable for the U.S. Navy's deployment. It began with those thousands of parts that Bob Masters was buying for him. Yes, for him, because he took his job very seriously and very personally. To deal with its complexity, work was carefully partitioned, planned in great detail, organized for maximum efficiency, and held to the strictest possible timetable. Anything that upset this carefully orchestrated concert of assembly was sure to upset Ralph Brown as well. Once final drawings were released to production, Engineers were not welcome on the factory floor, unless of course, there was a problem that only they could resolve. A missing item from a kit of parts needed for an assembly operation was the surest way to raise Brown's temperature. It was fortuitous that Bob Masters was sufficiently thick-skinned to take all that Brown could give, and then return the favor with a blast of his own for parts damaged or lost on the line and necessitating rebuys. The two were a good team, and also long-time friends, although that would be difficult to discern from outward appearances.

Brown was cautious but not ill at ease. Twenty-five years in manufacturing management had imposed a steady diet of status reports and management reviews. Brown knew what was expected, boiled the seemingly infinite quantity of production data down to its essentials, and presented it, bad news along with the good, in a matter-of-fact style. Jack Decker had not met Ralph Brown before Monday, but he felt as though he knew him well. There had been Ralph Browns before in Decker's career, and he had worked with them successfully by treating them with respect and dignity. The interpersonal skills were often set aside for a more direct approach. Hard core, hands-on leaders like Ralph Brown did not want to be bothered by "those idiots in management," so it was essential to be informed, supportive, and decisive. "No bullshit," as Brown would have put it. Jack Decker the generalist could wear that hat as well as his others. Conversely, in observing Decker during the day's reviews, during which he had said little, Brown had formed a positive first impression of Decker – at least compared to other Program Managers he had known – also unwelcome on the factory floor.

Brown got directly to the point of his presentation, the unsolved problems. "As you can imagine, Jack, we have had plenty of opportunities to screw things up on this program, and the remarkable thing is that we haven't done it completely, but we're still working on it. Bob Masters has already told you about the top thirty-two material problems, so I don't need to repeat that. What you need to know is that most of those shortages affect the satellite schedule, and until we get the parts, we can't finish the hardware. I have about twenty-five specialists cleared to work on the satellites in the clean rooms, and I'm running out of work for them while waiting for good parts.

"There are also some non-satellite lines affected by design problems. At the top of that worry list are four printed circuit boards for the surface ship platforms. They are dead in the water, so to speak, because they can't get through unit test. Engineering is working those problems, but until we have them corrected, some of my trained assemblers are not productively occupied. I'll have to release those people to other programs as early as next week, but there's no guarantee I'll get them back later. I may have to take the time later to train other people to do that work."

Decker asked engineer Lee Chang for his assessment of the engineering support to manufacturing. Chang's response was that the principal designers for those PC boards had been tasked to resolve those issues. "We think the basic design is OK. The engineering prototypes and the production first articles work correctly," Chang said, "but there appear to be some subtle differences in the production units. At this moment, I don't have a good forecast for any of the four."

Decker was not satisfied with Chang's answer, but he realized that it was not possible to forecast engineering breakthroughs with certainty. There was no point in pressing the issue further today, but he would do so tomorrow with the Chief Engineer. He did have a question for Brown. "You said these are for surface ship equipment. Are these same boards used on other platforms?"

"Yes," Brown responded. "The first of the boards through the line go to the ships because of the earlier need dates, but they're also used in the submarines, shore stations, and aircraft. Pretty soon those schedules will be affected, too. We're talking about a large quantity -- four separate boards used in a hundred thirty deliverable end-items all together. That's five hundred and twenty PC boards not counting spares. The first lot of half of the bare PC boards has been received from the vendor, about half of those have been through automatic component insertion, the critical hand-inserted components have also been installed, and ten of each type have been through flow solder. Those are the ones that don't pass unit test."

Decker suppressed his alarm. The line would have to be stopped until the problems were solved. Otherwise, additional defective units would be built and possibly discarded as very expensive waste. He said to Brown, "Ralph, it seems to me that you should stop that line until we have a solution. What do you think?"

Brown replied, "Well, that's the sensible thing to do, but I didn't want to jeopardize the schedule without some management support, which I don't have at the moment."

"You have it now. Stop the line until further notice. No point in building scrap. And Bob," addressing Masters, "hold up the vendor and delivery of the rest of the PC boards until we understand what's going on." Decker's firm decision and clear direction was what Brown had hoped for.

Brown's next viewgraph showed the net effect of the problems he had highlighted. The first satellite, scheduled for December delivery, would be a month late, possibly more. The surface ship communications equipment affected by those four failing microwave boards was already a month behind schedule with no reliable projection available. "It's a pretty small percentage of all the hardware we're building, but Murphy is at it again. As luck would have it, these problems are pacing both the satellites and the ships. Sorry to end on such a negative note, Jack, but that's where we are."

Decker said, "Thanks, Ralph. I appreciate your very candid appraisal." And then to the group as a whole, "Well, we can't just stand around and let these schedules slip, can we? I realize that you are all working very hard, but I think it's time to get some extra help. I'm going to work personally with Bob Masters on the parts problems. I'll show up early at your office where you have all the data first thing tomorrow, Bob. Let's try a few things that have worked before. The engineering problems are another matter. Maybe we need to get some graybeards working on them. Let me work that with Stan Abrahams. Matter of fact, Ralph, I'll set up a meeting with him tomorrow, and I'd like to have you and Lee attend. I'll let you know when." Decker's reference to graybeards, as they were respectfully called, indicated that he intended to have Chief Engineer Abrahams assign senior engineers with broad experience who could bring fresh approaches to solving the problems.

Katsu Yamoto, Manager of Product Assurance for SEANET, presented the third part of the production-related reviews. As he had indicated in the Monday meeting, Decker was convinced that quality could not be achieved by inspection alone. The modern approach to quality involved prevention rather than detection, and Decker was hoping to hear that philosophy expressed by Yamoto. He was not disappointed. Yamoto made that clear with his first viewgraph. He pointed out several vendor quality problems, but ended with "So on balance, I think that the SEANET quality program is working rather well. Best of all, the cost of quality, especially rework and scrap in the factory, is way down compared to the old days."

Decker said, "Well, I'm inclined to agree, Katsu. Time will tell, of course, but right off hand I can't see the need for any changes. However, I would like to have a daily report on the five vendor parts and the four PC boards that are in trouble. Just send me a brief email at the end of each day."

They proceeded directly to the Logistics Support presentation by Billy Barnes. There were no surprises here, and Barnes reported, "no serious problems, not yet at least since the logistics support activity up to this point in the program has been mostly analysis and planning and the development of support documentation. By the way, there won't be any of those three hundred page hard copy technical manuals with their awkward foldout diagrams. Everything maintenance personnel need will be on CD ROM for instant access on their laptop computers."

Barnes emphasized that his group had analyzed the design thoroughly to determine the level of spare parts, special tools and test equipment, and personnel resources that would be required to support the system during installation, test, and the first two years of operation.

"Well, Mister Barnes, I'm a little disappointed that you didn't have any unmanageable problems for us today as did most of your colleagues. But rest assured, your day will come soon enough. I'll be there with you, as will we all. Good presentation, Billy, and I appreciate it." In turn, Decker's lighthearted compliment was appreciated greatly by Billy Barnes.

In Decker's view the highest risk phase of the program was system integration and test. His review of the current program plan with Ken Martin had indicated that the schedule was unrealistic. Replanning and rescheduling were essential, and the Navy would have to be persuaded next week that it was in their best interest to do so. It was in this phase, he believed, that he could have the most impact on the success of the program.

The group settled in after the welcome afternoon break, and Decker resumed the meeting with a brief speech. "This has been a long day, and in my opinion a very productive one. Now, finally, we come to the end game, system integration and test. To borrow a phrase or two, this is 'where the rubber meets the road.' Or maybe you prefer 'It ain't over 'Tell the fat lady sings.' My personal favorite is Yogi Berra's 'It ain't over 'till it's over."

The team members smiled and relaxed a bit, pleased to note that Decker had not lost his sense of humor and was not shaken in spite of the serious problems presented to him. "The point I want to make is that whatever your job on this program, no matter how clever you have been or how hard you have worked, your job is not finished until we turn over the SEANET key to the Navy. So please give Tony Grazio your undivided attention as he walks us through the plan for demonstrating to the Navy that we have delivered what we signed up to deliver, and that they can proceed with confidence into their own operational test and evaluation prior to deployment of the system in the fleet. Tony ..."

Grazio projected a top-level summary of the system integration and test plan. In logical steps the system would be brought together piece by piece through increasing levels of complexity until the entire system could be tested to verify its performance as a whole. Step one in that evolution was the construction of a Land Based Test Site on the DSD grounds in which the first representative set of production hardware and validated software could be brought together and tested prior to shipment to the first Navy platform of each type.

Once the first full set of production hardware was installed and checked out in the LBTS, including a simulated space environment for the satellites, that equipment was designated contractually as the First Article. The contractor would work with it until satisfied that it was functioning as designed, and then a Navy-approved set of formal First Article Tests, witnessed by Government representatives, would be performed. When successfully completed, that contractual obligation was met, and the First Article was "sold off," in the jargon of the trade.

After first article tests were completed, every subsequent set of equipment would be installed and "sold off" in the LBTS prior to shipment to its platform for installation. Grazio projected a detailed schedule of installations in the one hundred thirty platforms -- twenty-five ships, two submarines, one hundred aircraft, and three shore stations. Billy Barnes' field engineers would rotate to the shipyards and air stations on a precise schedule coordinated with the Navy to install, check out, and test the equipment. Navy personnel from each location, having been trained at DSD, would assist them and take over the operation and maintenance of the equipment upon the departure of the field engineers. DSD logistics would remain on call full time to support the sailors if needed. The three satellites would be launched sequentially during the installation period from Vandenberg Air Force Base near Lompoc, California, headquarters of the Defense Department's Western Test Range.

When all platform installations were complete and the three satellites were in geosynchronous orbit, contractor system tests could begin, and at the end of successful contractor tests, the Navy's Operational Test and Evaluation Force, OPTEVFOR, would conduct independent and objective Navy tests to determine the suitability of the system for fleet-wide deployment. DSD could observe those tests, but was not allowed to provide assistance unless specifically requested by OPTEVFOR.

That was the broad outline of the SEANET System Test Plan. The STP, as it was called, was a contractually required data item developed jointly with the Navy and approved by them for implementation by the participating agencies. It could be modified only by specific contractual agreement between DSD and the Navy, and the Navy would need concurrence from other affected Government agencies.

"Tony, that's an excellent plan." Decker asserted. "I like it not only because it's textbook planning, it also matches my own experience. Nevertheless, based on the current program status as we have dissected it today, I'm sorry to say it won't work. If he were here today, Arnie Tell would say, 'Give us enough time and enough money, and we can put a man on the moon.' Well, we don't have enough time, and we don't have enough money. Therefore, we have to find a way to streamline this plan without introducing unacceptable risk. You've laid out what we ought to do, and I agree with you, but in the current circumstance we have to identify what we must do and what we can eliminate or simplify. I picked up an expression from a Navy test director years ago, and I wish I could remember his name so I could properly attribute it to him. His philosophy was 'build a little, test a little, and leap-frog when you can.' If we can find those 'leap-frog' opportunities, we have to take them. So let's go back to page one, and start over."

For the next two hours Decker, supported earnestly by Martin, questioned every step of the System Test Plan Grazio had so painstakingly developed. Others joined the discussion when activities for which they were responsible were explored. Schedule compression was proposed for common equipment since many tests could be abbreviated or eliminated once the first item of a type was verified. Very significantly, a number of tests were identified as redundant since they would have been performed previously in an earlier phase of integration. As he had hoped, Decker's borrowed expression, "build a little, test a little, and leapfrog when you can," became the guiding tactic for proposing a faster and leaner Land Based Test Site phase without introducing unmanageable risk. And while the specific dates for installation of equipment in the Navy platforms and stations was driven mostly by their availability, it was agreed that the elapsed time scheduled for those installations could be reduced by almost half with little risk. That would allow delaying the start of installations without impacting their end dates and would provide some relief for production equipment delivery dates. Another decision allowed some, but not all, tests currently planned to verify subsystem requirements to be deferred and verified when full system tests were performed.

Finally the discussion centered on the final contractor system test phase – the critical tests that required the entire system to be installed and all of its elements functioning. The group concluded that none of the system tests could be eliminated since they must remain comprehensive, especially since some lower level requirements were now to be verified at the full system level of integration. Decker agreed, but he opened the door to consideration of how and in what order the tests might be performed.

Milt Karinski suggested, "The long pole in the tent is still the satellite delivery schedule. Can we change the order of tests and do those tests not requiring communication with the satellites first?" There were, in fact, a number of tests not dependent on the satellites, and by bringing them forward in the sequence, the totality of system tests could be completed at an earlier date because it was not necessary to wait for the satellites before system tests could begin.

Tony Grazio argued that the test schedule had been arranged to minimize cost, and changing that arrangement would result in inefficiencies for both the contractor and the Government.

Responding to the debate, Decker offered, without condescension, another brief lesson in program management. "This is not new news for any of you, but sometimes we have to revisit the obvious. We are constantly confronted with tradeoffs between cost and schedule. I reminded you earlier that a shorter schedule is usually lower cost, all things considered, but not always. Usually we try to keep the cost down so long as we can still meet the schedule. In this case, meeting the schedule seems to me more important than the cost of testing. It may be that the elimination of earlier tests we are considering will offset the higher cost of a new system test schedule. We need to confirm that with some analysis, but for now let's go with shortening the schedule." The matter was settled.

Decker had been waiting for someone else to raise an important but controversial issue. No one did, so at this late hour he would have to raise it himself. The reluctance to raise the issue was due to the satellite launch schedule being considered sacred. Extensive coordination between a number of Government agencies was required, and that schedule had been arranged so that that all of the contractor system tests could be conducted with three satellites. Nonetheless, Decker asked, "Why do we need three satellites for the entire test program?

Could we do some of the contractor tests with just one then two satellites and launch the third near the end for just those tests requiring three?"

Martin responded, "It would certainly help, Jack, but The Navy and the Air Force and everyone else involved in the program would have to do a lot of replanning to make that happen. They would really be pissed! And we would need a contract modification. Do we really want to open that can of worms?"

"I don't like to do it, but we may have to inconvenience the customer and make him do a little extra work if it saves the program. I think the Admiral will go along and support us in this if it really helps, so let's take a look at it and see what the result would be." Decker was not as sure of the Admiral's support as he had indicated, but changing the satellite launch schedule had to be considered. The discussion that followed resulted in a consensus that the satellite launch dates could be delayed until actually needed to support the tests, and those dates could be met with much less risk than the current plan required.

"It's almost eight o'clock," Decker observed, "and I don't know about you, but I'm a little beat. I think we've got enough to make a new plan, so let's get it into the computer for a new schedule. What I would like you all to do is get your replanning inputs into Steve Martin's Program Control folks no later than close of business tomorrow. They can have a first run ready for us by Saturday afternoon, then we can fine-tune it to make a plan to show the Navy next week. You'll have your replan schedules on Monday morning so that you can get your presentations ready for the customer meetings starting on Tuesday.

"I know this twelve hour day has been rough on you, and there may be more like it needed to get through next week, so hang in there." Decker then instructed Martin, "Ken would you please review the action items you've been so faithfully chronicling all day?"

Ken Martin had, indeed, been taking extensive notes at the meeting and he went over his notes and assigned specific actions to each member of the program team, including Decker himself. Completing that in about fifteen minutes, he turned the meeting back to Decker.

Decker concluded the meeting with a note of appreciation. "Thanks for all your good ideas today. I'm damned well pleased with what we've accomplished, and I can't wait to see the results. I could not have asked for more. Get some sleep, and we'll hit it again tomorrow."

Like Decker, they were all pleased with what had been accomplished and also grateful to have Jack Decker at the helm. Now it was up to the team leaders, the Program Control group and Ken Martin to produce a revised program plan for presentation to the Navy in just four days.

CHAPTER ELEVEN
THE ASSISTANT

Decker arrived at his VP office at seven o'clock Friday morning. There was a neatly penned note on his desk.

8/20/00

Hello Mister Decker,

My name is Erik Hansen. Betty Emery asked me to assist you until you make a permanent arrangement. I understand you're tied up all day today, but please give me a call tomorrow at extension 5319 after about 7:30.

I took the liberty of going through the accumulation of incoming mail. Nothing urgent.

A few things require your signature or personal attention, and they're in your in-basket.

Regards,

Erik

"Well, I'll be damned," Decker thought to himself. "A male assistant. Well, Doctor Johnson had his Boswell, so I suppose I can make do with Mister Hansen for a few days." He would call Hansen as soon as he made two more-pressing calls. He needed to have several meetings scheduled, and Hansen could take care of that. First order of personal business was the promised visit to Bob Masters to get the data he needed on the subcontractors.

The first call was to Arnie Tell's office. He and Susan Anders had not yet arrived, so he left a voice mail message asking Arnie to return the call when he had a minute, adding, "It's important."

The second call was to Admiral Sullivan. Jane Wilson answered and put the call straight through to the Admiral. "Good Morning, Jack. You beat me to it today. How goes?"

"Good Morning, Admiral," Decker replied. "Things went rather well yesterday in our marathon brain storming session. We have the makings of a better program plan, and I believe it will solve most of our schedule problems. We also have some potential solutions for the more serious technical problems. I won't get into the details now, we'll do that next week with Commander Alvarez and your people, but I did want to give you a heads up on one important issue that needs your personal attention."

"OK, how can I help?" Sullivan asked.

"We need to reschedule the satellite launches," Decker stated succinctly and with conviction.

There was a lengthy pause, then Sullivan's response. "Well now, that's enough to clear my sinuses. Nothing like a good whiff of ammonia when you're dizzy. Did I hear you correctly? You want to reschedule the satellite launches?"

"We have to do it, Admiral." Decker wanted to sound dead serious and he did. "There's no point in kidding ourselves. The bad news is -- the satellites are going to be late by six weeks best case, eight weeks worst case. The good news is – it doesn't matter. At least it doesn't matter as far as supporting the overall system test program is concerned. By reorganizing the tests, we can launch one satellite per month starting in February and still get everything done well before OPTEVFOR takes over. We're working up the detailed plan over the weekend, and we'll get your team's inputs next week. But one thing is a certainty; we need more time to finish the satellites, so a big-time workaround is needed. I know this is a bitter pill for you to swallow, but that's the reality of the situation."

"Have you discussed this with Arnie Tell?" the Admiral wanted to know.

"No, sir. I placed a call to him just before calling you, but he isn't in yet," Decker answered.

"OK, Jack. Message received and understood, but I need to know a lot more before I can approve it. I'll rely on Jose and his team to be certain that your plan is as sound as you say it is. But Christ! The politics! That's why I need to talk to Arnie. Ask him to call me right away when you reach him. And by the way, tell him I said it's OK to let you in on our little secret. He'll know what I mean, and in view of what you've just told me, I think you have a need to know." The Admiral's reaction was about as good as Decker could have wished.

"One more thing, Admiral," Decker continued. "I'd like to call you at the beginning of the Tuesday Morning meeting. I'll put you on the speakerphone. I suggest you make a short speech informing your team on how important it is to come up with a program plan that we can all have confidence in, and that we may have to do some unconventional things to get where we need to go on time. Ask them to give us an open-minded hearing and then to help us make the best damn plan we can by working together. And I suggest you mention security again for both your guys and mine."

"Good idea, Jack," the Admiral agreed. "I've got your speech on tape, so I'll just play that back to them." They both laughed aloud and then said their good-byes.

Decker cleared the in-basket, signing all but one of the documents Hansen had left for signature and scanning the others. To one item he attached a note to Hansen.

8/21/00

    Erik,
    What's this?
    JD

It was almost seven thirty, so he dialed 5319. "Erik Hansen," the party answered.

"Hi, Erik. This is Jack Decker. I got your note, and thanks for going through the mail. Can you come over now?" Decker requested.

"Certainly, Mister Decker. I'll be there in five minutes."

"Come on in, Erik," Decker looked up and greeted him as Hansen knocked lightly from the doorway. "Very nice to meet you," he said as he stood and extended his hand.

Erik Hansen was a well-built, clean-cut, nicely dressed young man with abundant blonde hair and striking blue eyes. Decker estimated: Scandinavian, about thirty, six feet, one-eighty pounds, probably works out. Hansen had a grand smile, a friendly voice with just the slightest hint of a Northern European accent, and a pleasant demeanor both natural and unaffected. Decker remembered a favorite quotation, the source also forgotten long ago, "You don't get a second chance to make a good first impression." Making a good first impression was a problem Erik Hansen did not have.

Decker offered Hansen a seat, and they chatted briefly about their backgrounds. Hansen had come to the States from Norway with his parents when he was fifteen, and had become a naturalized citizen just before being employed by Tellonics in Burbank seven years ago as a mail clerk. He was transferred to DSD five years ago after receiving his Secret security clearance, and had held increasingly responsible administrative positions. Currently he was in charge of maintaining personnel records for Human Relations, thus his association with Betty Emery and his availability to assist Decker temporarily. He mentioned that he had been responsible for the introduction of personnel database software that had simplified record keeping dramatically. Decker exposed only that he had been a classmate of Arnold Tell, had worked for DSD previously and had been asked to return for the specific purpose of managing the SEANET program. So that there would be no misunderstanding about relationships, he informed Hansen that he reported directly to CEO Arnold Tell, not to DSD's Larry Hamilton.

They outlined Hansen's duties, Hansen volunteering to take on more tasks than Decker would have assigned. Hansen was taken aback at the discovery that Decker had two offices, this one for official "vice president stuff," as Decker put it, and the other for "my day job." They worked out a scheme for communication. Hansen would be stationed full time at Decker's assistant's desk with its multiple phone lines, intercom, and Intranet capability. To a caller, the existence of two offices would be transparent, including the daily call from Admiral Sullivan. Hansen would also have access to Decker's email and would filter it for all but that which Decker should see personally. With regard to filtering paper or electronic mail, they adopted the rule, "when in doubt, don't." Decker's routine would be to stop in the VP office first thing in the morning and last thing at night to take care of paperwork. During the day they would be in touch by telephone as the need arose.

For his very first assignment, Decker asked Hansen to set up the meetings that he had committed to during yesterday's planning sessions, and then he walked to his working office. Just as he arrived there at eight o'clock, the intercom line rang. Hansen announced, "Mr. Tell is returning your call on line one. Miss Anders is on the line." Decker noted the reference to "Miss" rather than the politically correct "Mizz" Anders.

"Morning, Susan," Decker said.

"Hi, Jack. Long time no chat. Here's Arnie," she replied.

Tell's opening salvo predictably included military slang. "What the hell were you doing up at oh-dark-hundred? Trying to make an impression? You think I'm supposed to drop everything and call you every time you say it's important? Dammit, if it isn't important you shouldn't be calling me. And where the hell have you been hiding all week?"

Having been put on the defensive immediately, as Arnold Tell's good-natured strategy demanded, Decker responded in kind. "You know me, Arnie. I just go to meetings so I don't have to do any real work. But we had one helluva meeting yesterday." He summarized the meeting and its outcome briefly, ending with, "So we decided to reschedule the satellite launches."

"Jesus Christ! That should go over big with the customer," Tell observed.

"Well, I told Rusty Sullivan, and he's at least willing to listen to reason. He wants you to call him to discuss the potential political fallout, and he said something about your sharing a secret with me," Decker said.

"OK, I'll call him right away before the sonovabitch takes off at noon for his usual golfing weekend at taxpayer expense." Tell could not resist an occasional impertinence at the expense of the Admiral. "As for the secret, I agree you ought to know about it, but it's something I can't discuss on the phone. I'll be down to DSD for a meeting Monday afternoon, and we can talk then."

Decker, as he was expected to do, and with immunity, took an irreverent jab at Tell. "OK, Arnie. I'll see if I can work you into my busy schedule. Just make an appointment with my office. And I have some intrigue to pass along that I think you will find very interesting." Now they were both in suspense, a calculated balancing of the scales on Decker's part.

After the Tell call, Decker dropped into Ken Martin's office to say hello and check on the status of replanning activity, and then he proceeded to Bob Masters office as promised. Masters and his staff had prepared single sheet summaries of the status of each of the thirty-nine subcontracts including the names, phone numbers, and email addresses of the general manager and sales manager, or equivalent titles, which varied from company to company. It was precisely the information that

Decker needed. He reviewed each of the sheets with Masters for background that either he requested or Masters volunteered as relevant. Decker informed Masters that he would be calling the executives of those subcontractors who had either schedule or quality issues to see what might be done and that he planned to talk to as many as time would allow today if they could be reached and to the remainder Monday. Masters said that all of the subcontractors had been notified of Decker's appointment as SEANET Program Director and that they would not be surprised to hear from him. They spent the most time on the Commware subcontract since it had the greatest impact, but Decker decided to delay that call until after he met with Arnie Tell on Monday.

Erik Hansen completed the phone calls setting up meetings as Decker had requested. He left a voice mail message for Decker notifying him of the meetings and also typed a short email note to Decker:

8/21/00 Jack Decker Appointment Schedule

11:00   Meeting in L. Hamilton's office with S. Abrahams
            Agenda: SEANET Software Priority
            Ada and C++ engineers for translator study
             Engineering staffing general

12:30   Scheduled phonecon with J. Talbot at Zybyte

13:00   Meeting in L. Hamilton's office with S. Abrahams, R. Brown, and L. Chang
            Agenda: Graybeards for microwave board production problem

14:00   Unscheduled calls to West Coast subcontractors

Hansen intended to suggest that he set up Decker's calendar on their computers so that Decker could plan his days more effectively. He would also inquire if Decker used a PDA, a Personal Digital Assistant. If so, he would set up a link allowing Decker to download the latest computer version of his calendar and to upload any manual entries he made so that the two would always be synchronized. If Decker did not have a PDA, he would suggest it and have one purchased for him. He understood that Decker's schedule would become very busy in the coming weeks and months, and he would need to organize his time carefully. "What did we ever do without computers?" Hansen thought to himself.

Decker returned to the working office and checked his voice mail, making a mental note of the two meetings Hansen had scheduled with Larry Hamilton, and then he checked his email, pleasantly surprised to see the schedule in written form. He printed a copy and placed it in the letter size notepad cum briefcase that he habitually carried with him throughout the day. There were several other messages that Hansen had left on the computer for Decker's information or required action after filtering out the "junk mail" as Decker had defined it in their earlier meeting. Included was an email from Hansen informing him that he had set up an address book on Decker's computers, accessible from either office or his laptop, containing email addresses and phone numbers of people Decker might need to contact in or out of Tellonics. Hansen would keep this up-to-date as communications demanded, and Decker could add to it also. The email included instructions on how to access the Tellonics computerized phone book that listed all Tellonics employees and offices worldwide.

Decker appreciated Hansen's initiative and would tell him so when he saw him next. He said to himself, "My Boswell seems to be pretty much on the ball."

The meeting with Larry Hamilton and Stan Abrahams started promptly at eleven. They exchanged greetings and small talk about Decker's apartment hunting prior to addressing their real purpose. Decker, having called the meeting, began with a brief summary of yesterday's all-day meeting and pointed out that the three topics on this meeting's agenda had come up during those discussions.

"The first issue is that SEANET does not appear to have a very high priority with regard to assignment of software engineers. I think Mary Stevens has done a super job under the circumstances, but she needs help, and we have to give it to her."

Stan Abrahams responded, "Yes, we've had a terrible time hiring and retaining qualified software engineers, but that turned around in the past several months. At one time we were about two hundred shy of what we needed, but that number is now down to fifty. I haven't heard any specific complaints from Mary lately. Do you know what she needs?"

"Not exactly," Decker admitted. "Let me suggest that you get with her directly for her input, and of course, her boss should participate. But then, still short fifty engineers, can SEANET get top priority for the next couple of months?"

"That's up to Larry," Abrahams replied. "He sets the priorities."

"How about it, Larry?" Decker asked.

"I'll take a look at it and see what we can do in view of other programs vying for those same resources."

Decker had not anticipated Hamilton's position. They might be at odds on this issue, so it would be necessary for Decker to take strong action in order to prevail. It was not at all unusual for a program manager to be in disagreement with the functional departments when there were resource conflicts. Disagreements were part of the business, and over time one learned how to "disagree without being disagreeable" as the saying goes. Decker, Hamilton, and Abrahams would not be in their current executive positions if they did not understand this. At times the arguments could become quite heated, but never personally offensive, because in all probability they would have to work together for years to come. Decker had another favorite quotation sans attribution: "Friends come and go, but enemies tend to accumulate."

Decker said, politely but quite directly, to Hamilton, "In other words, Larry, apparently you do not consider SEANET your top priority program without 'taking a look'. If that's the case, you and I need to have a conversation with Arnie Tell. When I took this job, Arnie told me that SEANET was his top priority, so if there's any misunderstanding about that, we need to get it straightened out."

"Well, of course SEANET is very high priority, but since you were in my job previously, Jack, you must understand that I have to take a helluva lot of things into consideration when assigning priorities – not the least of which is DSD's bottom line, for which I am personally accountable." Hamilton's point was well taken, Decker thought, but not defensible in this circumstance.

Decker got directly to his point and made it bluntly. "Yes, the bottom line is important, Larry, and I'll do everything I can to help you make your numbers, but it is not the primary consideration in this case. The Navy is not stupid, you know. If they suspect for a minute that we're diverting resources to fixed price programs at the expense of cost reimbursement programs in order to improve the bottom line, they'll have our collective asses. And I certainly hope it isn't the case, because if it is, somebody could get fired or go to jail -- and I'm not volunteering."

"Jack," Hamilton responded, "you know me better than that. But I admit that it could have an awkward appearance." He wondered to himself if he was as clean as he intended to be. The pressures of the bottom line could certainly influence decisions that Monday morning quarterbacks could call into question on ethical if not legal grounds. Hamilton continued, "Tell you what. Arnie Tell will be here Monday afternoon. I'll pass the buck to him. If he says SEANET is number one, then it's number one. You can join us if you like."

"I like," said Decker.

Moving to the next agenda item, Decker said, "As I mentioned before, we have to help Commware find a solution to the satellite software speed problem, and Mary suggested that some sort of translator or cross-compiler could be developed to convert the Ada source code to C++ object code that runs faster. I don't know if it can be done, but we need to check it out. If it looks feasible, then we'll give it a shot. So Stan, we need a small team to study the idea for a few days -- over the weekend if possible. We also need to do some research to see if it's been done before. Mary recommended a couple of Ada experts and a couple of C++ experts. Maybe there are some guys or gals who are expert at both, and that would be even better. Can I assign that task to Engineering?"

"Let me work on it," Abrahams replied. "It sounds too good to be true, which probably means it is, but sure. I'll take that on personally and get you a report early in the week if not Monday."

"That's great, Stan," Decker said. "Once an engineer, always an engineer, right?"

" Yeah, I still like to get into it once in a while. I'm a software guy, you know. Started as an EE, but drifted into software over the years before getting into manglement."

145

Abrahams had, indeed, started as an electrical engineer when digital computers were still a mystery to most. It was not until some years later that software was recognized as a separate engineering specialty for which degrees were granted by engineering colleges. Abrahams believed that an EE had certain advantages in developing software due to a more thorough knowledge of how the host hardware worked, and it just might come in handy in this interesting study for SEANET.

"OK," Decker continued, "that covers the past and the present. My third topic is the future staffing of the program. When we reach agreement with the Navy on a new program plan and schedule, there will no doubt be some significant departures from the current plan. That means staffing will be different. We won't have the specific requirements for another week, but my quick look assessment says there will be some short-term negative impacts on both manufacturing and engineering. I wanted to be sure to give you, especially Larry, a heads up that we may have to drop some production people from SEANET to keep our productivity realization at an acceptable level."

Hamilton was surprised. "That is definitely not good news, Jack. If you have to drop very many, we could be faced with a layoff because all other production programs are fully staffed, and we'd have no place to put your excess people. I need as much notice as you can give me in order to orchestrate a layoff if one is needed – hopefully, it will not be. And we need to give Arnie Tell a heads up on this on Monday."

"Right," Decker agreed. "I hope it doesn't come to a layoff, but if it does, then it does."

Layoff is a dirty, six-letter word in industry, and especially in the aerospace industries where experience and company loyalty are vital to success. The humanitarian dimension is not easy to reconcile in the hearts of executives who have to make layoff decisions, but the greater good must be served. The principal responsibility of a corporate executive is to insure the continuity of the business in order that as many employees as possible remain gainfully employed and stockholder equity be preserved. The inefficiency of carrying employees with no value-added work to do could not be long tolerated without jeopardizing the enterprise. Critics would call for the executives to cut their own salaries in order to pay workers who would otherwise be laid off, but the math did not work out that way. On the contrary it could be argued that a more expensive executive might have avoided the problem in the first place. Whenever Hamilton heard this debate, and he had heard it many times, he would console himself with thoughts of the sports stars whose compensation was ten times his own. To be sure, there were excesses, but like most corporate executives, Larry Hamilton had started at the bottom of his profession and worked himself up by hard work and demonstrated capability. Hamilton had, himself, been laid off on two occasions as a young engineer, and each time a better position had been found. Notwithstanding the positive case that could be made, Hamilton would do everything he could to avoid the pain of a layoff.

The agenda of the first meeting with Larry Hamilton having been completed early, the three of them retreated to the executive dining room for lunch. Decker excused himself just before twelve thirty to return to his VP office for the scheduled phone call to Jerold Talbot at Zybyte.

As he walked to his office, Decker reflected on the growth of digital computer technology over a quarter century, in which Zybyte had been a late comer but was now number three behind Intel and AMD. When he graduated from Cal Tech in 1974 the vision of the personal computer as a household appliance existed only in the minds of a few entrepreneurial innovators, Apple Computer's Steve Jobs and others of similar insight. In fact, few personal computers existed, those handcrafted from elementary components by technically skilled experimenters, not unlike the radio amateurs who played such

147

an important role in developing that technology. The early microprocessors ran at a snail's pace by year 2000 standards – about five hundred thousand instructions per second. MegaHertz speeds, millions of instructions per second, followed quickly, microprocessor speeds doubling about every eighteen months as predicted by Moore's Law. Intel's Gordon Moore postulated that the density of microcircuits would double repeatedly at a periodic rate, which rate turned out to be about eighteen months. At the beginning of the twenty first century military qualified microprocessor chips ran at speeds up to about two hundred megaHertz, two hundred million instructions per second. Decker speculated that gigaHertz microprocessor chips, executing billions of instructions per second, would eventually become available. The speed was determined largely by the smallest dimension of the semiconductors that could be reliably imprinted on silicon, gallium arsenide, or other wafers -- the smaller the faster. The ultimate physical limit would be determined by the size of the very molecules of the materials used. He did a quick estimate in his head, concluding that about a hundred gigaHertz, a hundred billion instructions per second, would be the theoretical limit. The practical limit would probably be half or a quarter of that. Time would tell. He hoped that Zybyte was well down that path.

Protocol required that the prime contractor place the pre-arranged call to the supplier, thus avoiding any embarrassment to the prime's representative should there be a delay. Decker did not hold too much to protocol except for first encounters, and Jerry Talbot had been a long time associate. Nevertheless, and as instructed, Erik Hansen placed the call to Talbot as soon as he saw Decker approaching.

"Hey, Jerry, how you been?" Decker said when Hansen transferred the call to him. Without waiting for an answer, he continued, "I guess you know that I'm back at Tellonics and running the SEANET program."

Talbot was aware of Decker's new position and, as a supplier to SEANET and other Tellonics programs, was somewhat apprehensive about the purpose of Decker's call. "I'm doing well, Jack, and congratulations.   Or should it be condolences?"

"Too, early to tell, Jerry. But it's one helluva program, and I'm really excited about it. We have our problems, of course, and that's the reason for my call." Decker wanted to get directly to the point.

"How can I help, Jack?" Talbot offered.

"We need a faster processor for the SEANET satellite motherboards. We've got some sluggish Ada software running in your favorite competitor's processor at two hundred megahertz, and we need twice the speed.   Have you got something on the drawing board that's compatible with the brand X processor instruction set and runs like a gazelle?" Decker presented the challenge.

Talbot took up the challenge without hesitation. "We might be able to help. I'll have to check on current status. You probably know that microprocessor instruction sets are pretty standard these days, and software emulators can take care of small differences, if any. Speed may not be a problem. We're already working on gigaHertz processors and are close to production on five hundred megaHertz chips. We might be able to provide a few engineering samples for you to play with even though they may not be mil spec. But what the hell do I know? I'm just the boss. Can we arrange for our engineers to talk?"

"By all means, and don't worry about the military qualification. We can take care of that later," Decker said. "My guy is Milt Karinsky, extension 3425. Can you have someone call him today?"

"Bet your ass I can, Jack. I'll get someone to call right away to see what might be possible. And if you don't get a satisfactory answer, be sure to let me know personally."   In addition to helping out a business associate of many years standing, Talbot recognized the marketing benefit of having a highly regarded defense contractor as the first user of the MS-500x microprocessor chip, and in a satellite at that, where reliability was paramount.

The serious purpose of the call completed, Decker and Talbot spent ten minutes reminiscing, discussing current issues, and speculating about the future before bringing the call to its cordial end.

It was ten minutes before the scheduled one o'clock meeting with Larry Hamilton, the second of the day, so Erik Hansen took the opportunity to discuss the note that Decker had left with him on Friday for an explanation. The unsigned note, addressed to Decker, was in a regular company mail envelope and consisted of a single, brief sentence: "The answer to your problem is well documented." Hansen stated that he had no idea what the message meant or who sent it. He asked if Decker wanted to get security involved, checking for fingerprints and the like.

"Never mind, forget it. I got crank letters all the time back at Delaware Microwave. I was also the subject of a lot of graffiti scribbled on the men's room walls – women's room, too, for all I know. I didn't mind as long as they spelled my name right. It goes with the territory. And some of them were funny as hell; others a bit too close to the truth." Despite dismissing the note as unimportant for Hansen's benefit, Decker wondered what the message meant, who sent it, and why. He would be alert for a clue.

The one o'clock meeting in Larry Hamilton's office was delayed a few minutes for the arrival of Stan Abrahams. He apologized for being late with the explanation that he was setting up the team to look into the software translator they had talked about earlier in the day. Four good people had been identified and set to work, and he would join them as soon as this meeting was over. "I wouldn't want to miss any of your meetings, Jack," he said to Decker, "just in case you have more ridiculous action items for me." Decker and Hamilton, laughed, but Ralph Brown and Lee Chang were more reserved in their response. They were, after all, meeting with executives two or three levels higher, respectively, in the organization. But it soon became apparent that they were among men with robust senses of humor, an essential element of higher management credentials in Decker's view.

Decker responded with, "That's what the big bucks are for, Stan, being sure that everyone has something useful to do – especially the engineers." This time all four joined in the laughter, especially Brown, appreciative of the jab at engineering. "And that brings us to the point of this meeting. Ralph and the manufacturing folks are having a helluva time getting the production microwave cards to pass unit test, and so far Lee's people have not been able to find a solution. So I've put a hold on the production of those units until we get things sorted out. In yesterday's meeting we decided that we needed to call in the graybeards. Can you help us there, Stan?"

Abrahams asked Chang to describe the problem and the background in more detail, which he did – quite lucidly, Decker thought. Brown offered some relevant facts also to define the problem more specifically from the manufacturing point of view. Then Abrahams said, "You're probably right, Jack. We've got some super A-Team RF engineers, but naturally, they're in great demand. Let me see what I can do. Two guys probably. I'll have to get back to you."

"OK, then," Decker said, "end of meeting. If I had known it would be this easy, I would have just made a phone call. Sorry to take up your time, Larry, with something we could have handled independently. I just wanted you to be familiar with the big ticket items."

Hamilton was not at all upset. "I wish all the program managers would let me in on their dirty little secrets, Jack. Usually I find out about them after it's too late to do anything about them. So don't worry about wasting my time. Keep me in the loop ... please."

They went their separate ways at one twenty, which gave Decker an extra half hour for calls to subcontractors. Maybe he could get in a couple of east coast calls before two.

Erik Hansen had made copies of the one page subcontractor summaries provided to Decker by Bob Masters that morning, had made the appropriate entries in Decker's address book, and had checked for the availability of the identified executives for a possible call this afternoon from Jack Decker, the new SEANET Program Director. Hansen had identified himself as Jack Decker's personal assistant, and as a result of his carefully phrased inquiries and cordial manner, they had all agreed to be on the lookout for a possible call from Decker today or Monday and would have the appropriate members of their staffs available to participate in the conversation if necessary.

For the next three and a half hours, as Hansen got them on the line in turn, Decker talked with representatives of the subcontractors with either schedule or quality problems, including two on the east coast. In every case Decker received assurances that top management attention and high priority would be given to resolving those problems. From each company he requested and received commitments for daily email status reports on those issues until they were satisfactorily resolved.

Five o'clock. Friday afternoon. Tradition had it that this was the most likely time for a consuming new problem to arise just in time to spoil the weekend. With that in mind, Decker answered Erik Hansen's intercom call. "Would you like to try another call, or shall we call it a day? And by the way, Miss

Anders from Arnold Tell's office requested that you check with her before leaving for the day. Shall I get her on the line now?"

"Susan? I'll just call her directly. And no, no more calls today. I'm beat, to tell you the truth." Decker was a little surprised that he had been so open with Hansen. Maybe this was going to work out just fine. "Take the rest of the day off, Erik, compliments of the management. And I'll see you first thing Monday."

Hansen said, "Well, I know you'll be in over the weekend and may need some assistance, so I've already prepared my wife for my coming in this weekend if you need me. My home number is on the computer."

Decker had not realized that Hansen was married since it did not come up during their interview this morning. He replied, "No, I don't think it will be necessary, but thanks for the offer."

Decker dialed Arnold Tell's extension, and Susan Anders answered, "Working late again on a Friday Afternoon, I see. Just like old times."

"Yeah, not much has changed in the program management business," Decker replied. "I presume our boss is responsible for this call. How does he plan to interfere with my life now?"

"Well, not exactly Jack," she replied. "Arnie did ask me to check with you to see how your personal arrangements are coming along and to give you a hand if you need it. I understand you've been apartment hunting."

"Yes, and I've found a pretty good arrangement nearby. I expect to sign a lease sometime this weekend if I can find the time. It's a nicely furnished townhouse condominium near the plant. So I guess I'm pretty well set, Sue. But I do appreciate Arnie's offer." Decker was pleased that Arnie Tell was watching over him.

"Actually, Jack, it was more my idea than Arnie's. Don't get me wrong; he does care a great deal about your well being.

He considers you one of his few close friends, but he knows you can take care of yourself and doesn't want to interfere unless asked. But it occurred to me that you must be pretty tired of the DSD cafeteria and Pizza Hut. Could I interest you in a home cooked meal Saturday night? I'm having some friends over and thought you might like to join us." Sue Anders' invitation was well rehearsed. She had always considered Jack Decker a terrific guy, and now there was an opportunity to become better acquainted if he would permit it.

"I'd be delighted. What a nice thought, Sue. When and where?" Decker's acceptance of the offer was guardedly enthusiastic, since he, too, wanted to become better acquainted with Susan Anders if she would permit it.

She gave him directions to her Manhattan Beach house, her phone number in case he got lost, and ended the conversation with, "I'm so pleased that you can come. See you tomorrow night."

CHAPTER THIRTEEN
SATURDAY

High-level executives enjoy their weekends off just as much as anyone else, especially after putting in long days during the week. In Jack Decker's view, one's job was not the most important thing in life, but it came close. Dedication and willingness to sacrifice time off for the good of the enterprise were not only expected but were essential. It was important in this view to manage stress. Stressful events in his line of work were inevitable, but he handled them on the fly. He was not a worrier, and worrisome problems are the principal cause of stress. He was a planner, and he had developed a general methodology for managing problems that can be summarized simply. The best way to avoid problems is to anticipate them. If there is a problem, there is also a solution. Don't ignore problems, face up to them. Be certain to understand the problem clearly before trying to solve it. Get people involved, including your boss. Don't be afraid to ask for help. Never lie or mislead. It is counterproductive to get upset or to appear to be upset. This final point of Decker's view was the key to his management style. He had been criticized in the past for being "too laid back" or "liaises faire," but it was hard to quarrel with his record of achievement. He didn't intend to change now, so occasionally working on a Saturday or a Sunday was not a big deal. Neither did he hesitate to ask others to do the same, especially the higher paid professionals, but he did so sparsely and only when truly important. This weekend was important.

He arrived at DSD at eight o'clock, a little later than usual, and checked in first at the VP office to clear the IN basket since he had been too tired to finish it when he left Friday night at seven. Then he went to Ken Martin's office to see how the replanning effort was progressing. Martin and Steve Murphy were examining computer printouts as he entered and greeted them, "Good morning. Hope you slept well."

"Well, but not very long," Murphy answered.

"Roger that!" Martin added. "We were here 'Tell ten last night, but I think we have most of the department inputs sorted out. There are just a few missing, and we should get them this morning."

"I suppose it's too early to see the full schedule, then. Anything I can do to help?" Decker offered.

Martin answered, "Well, I think we've got things pretty much under control, so there's nothing you can do just now. As soon as we get the rest of the input data, and that should be before ten o'clock, we'll make a computer run and see where we stand. We expect a fair number of disconnects and duplications and omissions, and it takes several iterations to get it right. It probably will not meet our objectives at first, so we'll have to negotiate changes with the departments. That's when I think you need to get involved personally."

"Right," Murphy commented. "Some of the inputs we've received so far look pretty good, but others don't show much improvement. Believe it or not, some are a bit worse. We have a way of comparing the new schedule and cost estimates with the current plan, and we can give you a roadmap of where to look for deviations from targets."

Decker quipped, "OK. I guess I know when I'm not wanted, so I'll just go have a beer somewhere else. I've been thrown out of better joints than this you know." Martin and Murphy smiled as Decker continued, "So how about one o'clock? I need to get out of here by five today, and that should give us a few hours to work with the departments. I have some personal business to attend to, so I'll just disappear off campus for a while."

Murphy said, "Yeah, we should have a reasonable cut by about one thirty if everyone comes through. We'll ask the department leads to hang around for 'negotiations,' as Ken put it."

"See you at one thirty then, boss," Martin added. That choice of the word, boss, was comforting to Decker, indicating as it did that he had been accepted by the members as leader of the SEANET team.

Decker stopped by his working office to confirm his appointment with the rental agency that he had been working with on the furnished townhouse. They had the paperwork ready and could see him at ten thirty, so he left DSD intending to have a leisurely breakfast, which would also suffice for lunch, and then he would go to the appointment with the rental agency. He had decided to take a six month lease on the townhouse but would try to negotiate a better price by paying the full amount in advance. The agent consulted the owner, and there was a brief negotiation. The owner accepted the lease with a ten percent price reduction, a good tradeoff against potential interest that might be earned on the advance.

Decker returned to Ken Martin's office at one thirty as promised. Martin and Steve Murphy were waiting for him there. Martin told Decker that he and Murphy had been over the schedule several times with the department leads whose schedules caused a departure from the desired end date for the total program. These critical path activities had been reduced to the point of maximum risk acceptable to the responsible departments, but there had been no arm-twisting on Martin's part to go beyond that. At this point there had been a two-month improvement over the current plan in the completion of contractor systems tests based on the assumption that the Navy would approve the proposed satellite delivery dates, but another three weeks improvement was needed. Murphy suggested they go to the conference room next door to join the three department leads who were still struggling with their schedules.

Before they went to the conference room, Decker told them he was, "... very impressed that you've made that much improvement. And I agree that we should not pressure the departments to go beyond what they consider acceptable risk. If you think you're at that point now, a different approach may be needed. Let's take a look at the items that are still causing the schedule problem and see if any of them can be reduced in scope or eliminated or deferred in such a way that the main objectives of the system tests can still be met. In other words, let's be sure we don't have the tail wagging the dog. Remember that the Navy is willing to change the details of the contract delivery schedule if it makes sense to do so. The one thing that they hold most sacred is the turnover date to OPTEVFOR. That's the date we have to meet."

Several of the Department leads were in the conference room looking over their latest schedule printouts. Decker suggested they go over the program level critical path together to see if one or more tasks might be simplified, shortened, or deferred. After considerable inspection and discussion it came to light that the tasks on the critical path were very much the same as before. Most schedules had been improved, but several had not been changed at all. Among those was the satellite software schedule. Milt Karinski and Mary Stevens were in agreement and were reluctant to shorten that schedule until they saw some light at the end of that tunnel.

Decker asked Karinski, "Did you hear from the engineers at Zybyte yesterday?"

"Yes," Karinski replied, "and I was encouraged by their progress on the X500 processor development. Their processors use a different socket, so they're sending us four samples with hardware adapters and some emulator software so that we can start testing on Monday."

"Suppose the new chips are twice as fast? Will that solve the problem," Decker prodded.

"Probably," was Karinski's guarded response.

"And Mary," Decker addressed Stevens, "what have you heard from the translator team?"

"They're just getting started, of course, but they do believe that the idea is technically feasible. I had a call from them just before noon with a number of questions, and I'll be joining them after we finish up here." Stevens sounded optimistic.

Decker believed it was time for a management adjustment. "Tell you what. We have two plausible parallel paths toward solving this problem, and each one is about a fifty-fifty proposition. That means we've got about a seventy five percent chance of a schedule improvement. With those odds, I'm willing to take some risk. Generally speaking I don't ask people to sign up to schedules they are uncomfortable with, but this seems to be an exception. Let's take two weeks off the software schedule, and look for another five days elsewhere. Any problem with that?"

With nothing heard, Ken Martin said, "So be it. And I'll work with the leads to take a day from each of five other tasks on the critical path to make it. After all, who can estimate a schedule within twenty four hours?"

"Looks like we've got a plan, folks," Decker said. "Let's get it into the computer and get ready for the Navy next week.

CHAPTER FOURTEEN
DINNER

Manhattan Beach is about 20 miles southwest of the center of Los Angeles, one of several popular and overcrowded beach communities along the Pacific shoreline. It is noted for its wide, concrete "boardwalk" separating the residences on the southwestern side of PCH, the Pacific Coast Highway, California 1, from the gently sloping, sandy beach. A favorite route of tanned roller-skaters, bikers, joggers, and strollers, it is a voyeur's paradise year-round.

Doctor Phillip Anders and his wife Elizabeth bought their weekend retreat there in 1950 along with the adjacent empty lots on either side, and it had been used by them and their children, Mark and Susan, as a home away from home ever since. Not a mansion by any measure, but with enlargement and improvement over the years, it was, nevertheless, an impressive house featuring a spacious living and dining area under a vaulted ceiling. The wall facing the ocean was contiguous floor to ceiling windows of tempered glass through which was presented a glorious view of the Pacific and, on a clear day, the Channel Islands. In this neighborhood, it was where "the Joneses" lived. Although she resided in Burbank near her Tellonics workplace to minimize commuting time, Susan Anders often spent weekends here, and she used this splendid spot to entertain friends from time to time. She had taken great pains to make this a special evening for her guests – especially Jack Decker.

The guest list included her older brother Mark, like their father a medical doctor, and his wife Jennifer and their childhood friend Avery Parsons, Ph.D, a professor of history at UCLA, and his wife Norma. Susan and Jack would be the third couple in the party of six. It should be a very compatible group, she had estimated, with sparkling conversation, possibly along intellectual lines. It would be a very nice evening with an opportunity for

Jack and Susan to become better acquainted in a relaxed atmosphere away from the office.

Like Jack Decker, Susan Anders had once been married. Shortly after graduation, she had married a classmate at UC Santa Barbara, where she had majored in English Literature. The marriage was a failure almost from the start. Her husband had taken to serious drinking, and he had become physically abusive. Unlike many battered women, she had close family ties and did not feel trapped, so she was able to walk away from the marriage relatively unscathed. Luckily, there were no children to complicate matters. Discarding her married name after the divorce, she began anew with her maiden name, Susan Anders. Although she had become distrustful of men as a result of having to separate from a man she once loved yet came to despise, she had dated from time to time, and had had two fairly stable affairs, neither very satisfactory. "Jack Decker? Well, we'll see," she thought.

Brother Mark and Norma arrived first, and Jack Decker shortly after that. After introductions and polite exchanges, "How were the directions?" Susan inquired.

"Well done, Sue, " Jack replied. "Got lost only once, and that was my own cockpit trouble. It's been a while since I navigated these waters." And then, taken aback by the ocean view, "My God, Sue, what a fantastic view of a Pacific sunset. I don't think I've ever seen anything quite like it. How on earth did you find this place?" Mark took the lead with an explanation of his and Susan's parents' acquisition and loving devotion to the "beach house," as they called it.

Shortly the Parsons, Avery and Norma, arrived, and after the customary introductions and polite, get-acquainted exchanges, "Can I get you all something to drink?" They placed their orders, and Mark and Susan set about preparing them.

Decker ordered, "Johnny Walker black, on the rocks, if you have it. If not, any old Scotch will do – the older the better, of course." Avery and Norma Parsons ordered a Manhattan and a glass of white wine, respectively. Mark announced as he poured that he was having, "JD and branch," signifying Jack Daniels Tennessee Whiskey and water, and Susan joined Norma with a glass of wine. The Chardonnay was an excellent but little known Napa Valley wine by Steele Vineyards, and Decker recognized the distinctive bottle at once. It was not an exact science, but Decker believed that something was revealed about a person by the cocktail he or she ordered. The wine was appropriate for the ladies, bourbon or scotch on the rocks or with water usually chosen by men with no serious hang-ups, if taken in moderation, but he had not heard a Manhattan ordered in some years. Signifying nothing, he calculated.

Susan had prepared shrimp and crab hors d'oeuvres that were not only delicious but could have been a meal in themselves. Decker exclaimed, "Well, I must say Sue, this is right up there with Pizza Hut!" Only Susan laughed at this, the others somewhat surprised until Susan revealed the source of the friendly jab, and then they all joined in.

"Which Pizza Hut do you recommend, Jack?" Norma Parsons asked with a smile. "Perhaps we can have our daughter's wedding reception there." The ice had been broken and the group proven compatible, as Susan had hoped, although Avery Parsons was less a participant than the others.

After a second round of cocktails, which Decker did not finish since he was driving alone and knew that wine would be served later, they were invited out onto the patio for dinner. It was a beautiful, warm, clear, August night, and the three-quarter moon gave off sufficient light to cause a fluorescent sparkle on the gentle breakers and to backlight the activity on the boardwalk between the distant streetlights up and down the beach. The table was set eloquently for six, and Decker was invited to take a seat at Susan Anders' right as she sat at one end of the table. Counter clockwise from Decker were Norma Parsons, Mark

Anders at the other end seat, Jennifer Anders, and Avery Parsons, directly facing Decker.

Every detail of the meal had been planned carefully and servings prepared in advance so that it was only necessary for Susan to retreat to the kitchen briefly between courses. Each course, in turn, was brought out on a rolling cart with six plates already served and ready for presentation to the diners. It was a very efficient way to serve a fine meal, as Decker was quick to observe.

"Sue, I'd like you to come down to Buena Park next week and show Ralph Brown how to set up an assembly line." They did not know Ralph Brown, of course, but it was clear that he must be a manufacturing expert, so they were amused by Decker's innovative compliment – except for Avery Parsons. Parsons had not restricted his alcohol intake and was beginning to participate in the conversation.

"Who is Ralph Brown?" Parsons asked.

"Just a guy in Manufacturing," Decker replied. "He may be the best manufacturing planner I have ever known, so I hope you will take my clumsy attempt at humor as a very positive compliment, Sue." Decker felt trouble coming, but he was determined that, no matter what, he would not spoil Sue Anders' evening.

Susan showed her support immediately, also sensing a negative situation developing. "Of course, Jack. I'm very flattered in fact. Buena Park may be the best defense plant in the country."

Brother Mark had been asked to uncork and serve the dinner wine, a French Burgundy of recent but very good vintage year, 1999, and it went well with the beef Burgoyne, freshly steamed asparagus tips served cold under a delicate white sauce, and miniature white potatoes halved and braised in extra virgin olive oil with Susan's selected spices. The desert course was strawberry shortcake, obviously freshly baked, with real, not canned or frozen, whipped cream topping. "The strawberries are from Pleasant Valley up in Ventura County, and this is just the right time of year for them. California's best, I'm told," Susan informed her guests. They were obviously pleased since not a bite was left on any plate.

"Let's have a brandy and coffee inside," Mark Anders suggested because the air had begun to chill with the sea breeze. They took seats in the comfortable conversation pit of the open living area, and on the center table between the luxurious sofas was an array of colorful brandy bottles: Grande Marnier, a rare vintage Courvosier cognac, Triple Sec, Drambouie, Tia Maria, and several fruit brandies. A large, pre-warmed brandy snifter and coffee mug were ready at every seat. Decker poured himself a scant half-inch of Drambouie, still keeping track of his alcohol intake for the drive home.

Avery Parsons poured a half-full snifter of Tia Maria and added a generous shot of cream. He had already had too much to drink, and the sweet brandy was sufficient to push him over the precipice of belligerency. Parsons had known the Anders family as next-door neighbors in Brentwood, the fashionable Los Angeles suburb. As a teenager, he had fallen in love with the girl next door, but she had never given the slightest indication of a romantic thought toward him. Eventually, they went their separate ways, but the feeling for her lingered. During the evening he had become extremely jealous of the attention Susan was paying to Jack Decker, and he decided to confront him. Parsons intended to draw Decker into a controversial conversation in which he, Parsons, would occupy the high moral ground.

"So you're a member of the military-industrial complex. Is that right, Jack?" Parsons asked with practiced innocence.

Decker did not want to get into this subject with Parsons, but felt obliged to respond politely. "Well, I suppose I am at that, now that you mention it. At least that's what President

Eisenhower called it," was Decker's carefully worded response. "Eisenhower's warning of a potential conspiracy involving the military and its industrial partners never came to pass, of course, so the phrase is seldom used these days. I tend to look at my job as just doing work for the Defense Department. But it's pretty dull stuff for most people, so why don't you tell us about your work at UCLA. I'm especially interested in how young engineers are being educated in the humanities."

Decker had been in this position before, challenged by an anti-military academe with a utopian worldview. Right or wrong, those who chose to challenge Decker's profession were in for a serious debate, because he was extremely well informed. As a widower for the past eleven years, he had read widely with a particular appetite for world history and geo-politics. He subscribed to and read opinion journals representing various points of view. And although quite capable of vigorous debate, he preferred to observe Mark Twain's admonition: better to keep your mouth shut and be thought a fool than to open it and remove all doubt.

"I suppose you need lots of young engineers to help build your weapons of war that kill innocent women and children," Parsons persisted. "How can you sleep with that on your conscience?"

Decker, a student of classical debate, could not let this pass. "That's a rhetorical question based on a faulty assumption. Since I challenge your assumption that I am involved in the killing of innocent women and children, I must ask you for proof of your assumption. If you cannot provide it, then this conversation is over and you have lost the argument before it begins. Regarding sleep, right or wrong I work my butt off to help protect this country from its enemies, so I usually get less sleep than I really need, but I have no trouble at all sleeping when the opportunity presents itself. My conscience is clear. Any more questions Avery, or shall we change the subject and just be friends?"

Parsons did not respond to Decker. Instead, he took his wife by the hand and said, "Well, dear, I think it's time for us to run along. Nice party, Sue and Mark, and thanks for everything. We'll have to get together at our place real soon."

The four of them stood speechless while the Parsons headed for the door. As they departed, Norma Parsons turned and said to Jack Decker, "It's been a pleasure, Jack, and I wish you the best of luck in your new position. 'Nite everybody!" And they left.

"I'm really sorry, Sue. I realize he's an old family friend, so I tried to change the subject. But I just couldn't let his accusation pass. I should have just ignored it. Sorry if I ruined your evening." Decker's apology was genuine and obviously so.

"Don't apologize," Susan said. "It's about time somebody told off that pompous ass. He is an old friend, but he does get out of line when he's had too much to drink. Besides, I really enjoyed your response – the highlight of the evening."

"Right on," Mark Anders concurred. "I'll see Avery soon and tell him what an ass he was. I expect there will be no hard feelings. And Norma is a real jewel. She must have been terribly embarrassed. Maybe you should call her tomorrow, Sue, and set her mind at ease. It's really no big deal, and you handled it extremely well, Jack."

"Well, thanks," Decker said, "but I'm still a little uncomfortable with having offended your guest. Next time I'll be a little gentler."

Next to leave were Mark and Jennifer. They said goodbye to Decker, and then aside to Susan as they stood in the doorway, Mark whispered, "I like your Mister Jack Decker very much. Seems like a really nice guy, and smart as hell, too."

Jennifer added, "I like him, too. Let me know if you get tired of him, Susan." With that, Susan Anders and Jack Decker were left alone in the beach house.

"Can I help with the dishes, Sue?" Decker asked.

"Why not," Susan replied. Actually, she had pretty well taken care of loading the dishes into the dishwasher as the evening progressed, so there was little left to do -- just a cleanup of the conversation pit with its brandy snifters and coffee cups. Not everything would fit into the dishwasher, so a few things had to be washed by hand. "I'll wash, you dry," Susan ordered.

"I wonder what Arnie Tell would say if he should walk in on this little domestic scene," Jack asked. They laughed aloud, imagining the event.

Jack thought Tell's initial volley would be something like, "What the hell are you doing up here acting like a bloody husband when you've got work to do at DSD? And you, Susan, you're not supposed to fraternize with the hired help."

Susan imagined Tell's response would be along the lines of, "What the hell are you two doing in the kitchen when there's a perfectly good bedroom available?"

The kitchen chores completed, Susan and Jack returned to the conversation pit for another cup of coffee. Each found it very natural to converse with the other, so they talked about many things, mostly business – things they had done since they had last worked together in 1989, people they had known before, and what lay ahead for them. Glancing at his wristwatch at half past eleven, Jack said, "It's getting late Sue, so I'd better be going."

Susan Anders considered carefully what she would say next. Realizing that Jack Decker was a true gentleman and unlikely to make the first move, she responded with, "Jack, you don't have to drive all the way back to Buena Park tonight. There's plenty of room here, so why not stay over. Have some breakfast and a walk on the beach tomorrow. Surely SEANET can spare you for one Sunday."

Jack was grateful that Susan had made the first gesture toward a personal relationship. Having spent this wonderful evening with her, it was his intention to do so himself, but not tonight. No reason to rush things, he believed. His response, however, also carefully considered, was, "Now there's an offer that can't be refused if I ever heard one."

Their first kiss was exploratory. The next, comfortable. The next, passionate. "My place or yours?" Jack asked.

"Mine, of course," Susan replied. "Right this way," she said, as she took him by the hand and led him to her bedroom.

****

Jack and Susan parted company Sunday afternoon following a very long walk on the beach and their first intimate conversation, each revealing to the other personal information seldom shared with others. They found it remarkable that after these many years they should have a personal relationship. Both realized this had not been a casual one-night stand, so they vowed to let their relationship mature slowly over time if fate so willed but without long-term commitment.

Office romances are generally frowned upon in the business world because they frequently lead to trouble of some sort, so precautions had to be taken. Jack made it clear that his job would require long hours, frequent travel, and little time off, but Susan understood completely as an observer of the higher echelons of corporate management. They would continue to see each other but discreetly, mostly on weekends, and they would keep their relationship to themselves for the time being. They would not discuss other than business matters on company premises, telephones, or computers. Their personal communications would be strictly "off line," as Jack had put it

Decker returned to Buena Park later in the afternoon. He stopped by Ken Martin's area to see if anyone from the SEANET program team was working and needed assistance. The only people present were Steve Murphy and four Program Control analysts. Decker met those four for the first time, spoke with each of them individually, and let them all know how much he appreciated their coming in on a Sunday to get the reports ready for the Department Leads on Monday morning. They were appreciative of a Senior VP taking an interest in their work, and they let him know that they would stay as late as necessary to complete everything needed for Monday morning. Decker left his Marriott phone number with Murphy in case he needed to contact him for anything. He could move into the Town House anytime now, but would not do so until he had arranged for a telephone there, possibly Monday or Tuesday.

CHAPTER FIFTEEN
ELEPHANTS

When high-level corporate executives get together, they are referred to as "the elephants." As the analogy goes, they stomp around, inadvertently crushing anything in their path, generally travel in circles, and leave an awful mess for the rest of the organization to clean up. On this particular Monday, CEO Arnold Tell would meet both together and separately with two of his Senior Corporate Vice Presidents, Larry Hamilton and Jack Decker. These three elephants had much to accomplish on that day.

As promised, the Program Control data were ready first thing Monday Morning. No doubt the Department Leads would find errors or anomalies in the data that would require correction, but the work was essentially complete and would be ready for meetings with the Navy on Tuesday. After checking with Ken Martin and Steve Murphy, Decker proceeded to his VP office to take care of paperwork and continue his telephone calls to critical suppliers. Erik Hansen was ready for him with several documents requiring his signature and with the day's schedule. The meeting with Arnie Tell and Larry Hamilton would take place in Hamilton's office at two o'clock. He would meet privately with Arnie in his own office at four o'clock after Tell's one-on-one with Hamilton. This was the order that Tell had requested, and it made sense to Decker. Tell and Hamilton had issues to cover beyond SEANET; so getting the common SEANET issues out of the way first was the logical thing to do. At the end of the day Tell was to disclose the secret information closely held with Admiral Sullivan, and Decker had important information to share with Tell that Hamilton did not have a need to know. Decker thought it all had the ring of a cold-war spy game.

All but one of the remaining twenty-three critical subcontractor CEO calls were completed before two o'clock. The exception was the call to Mason Crenshaw at Commware. Decker needed to talk to Arnie Tell before making that call. The others, to a man (actually, two were women), had been cooperative. They all knew that they were on or near the SEANET critical path, and they agreed to apply the priority and resources necessary to bring their delivery dates and technical performance as close as possible to the targets. Five CEO's had requested technical assistance from Tellonics, and Decker agreed to provide it. For those cases he now drafted and transmitted an email to Ken Martin and Milt Karinski directing them to provide that assistance with as much dispatch as program priorities would allow. He placed a copy of the email in his "Reminders" folder so that he would remember (or Hansen would bring to his attention) that follow-up on his part was required.

_____

"You're a minute late, Decker," Arnie Tell reprimanded Decker as he entered Hamilton's office for the two o'clock meeting.

Consulting his wristwatch, a Citizen "Blue Angels" pilot's chronometer that he knew to be accurate within a few seconds, Decker was able to retort, "Actually, Arnie, I'm forty-nine seconds early. When are you going to buy yourself a decent watch?"

"Screw you, Decker, I've got better things to do than advise the Naval Observatory on the bloody time of day – and so have you. So what the hell did you guys drag me down here for?" The routine monthly one-on-one with Larry Hamilton had been scheduled weeks before, as everyone knew, but they conceded the point realizing it was just harmless Arnie Tell contrariness.

"The most important thing, Arnie, is to establish some priorities," Hamilton offered. "Quite naturally, Jack wants me to give SEANET top priority for shared resources, but some of the other programs – more profitable ones – will suffer if I do that. I'm perfectly willing to give Jack the support he needs, but you need to know that there could be a bottom line impact. The second issue is a potential layoff. SEANET has some non-productive manufacturing labor assigned waiting for technical solutions or process corrections still in progress. We can't absorb Jack's excess people right now, so it's either overhead growth or layoff. I ...we ...wanted you to know about those impacts before proceeding."

Decker said nothing, the issues having been succinctly described by Hamilton, and they waited for a response from Tell. Tell put his elbows on the small conference table at which they were seated and rested his head in his hands while he stared at Hamilton and considered his response. "A two-hander," Decker thought to himself. As a twenty-five year observer of Arnie Tell's body language, he knew that resting the chin on one hand indicated full attention, while the Tell chin in two hands was a sign of deep contemplation.

"Couple of questions," Arnie finally said. "How much profit impact this quarter and next, and how many people for how long?"

"Two hundred K for the remainder of this quarter, and up to five hundred K next quarter. And the layoff looks like about forty to fifty people for three months." Experienced executive that he was, Hamilton had anticipated the questions and was ready with the answers.

"Peanuts," Tell replied. "Well, maybe walnuts, but I can offset the profit impact with some unplanned plusses on the commercial side. As for the layoff, 50 people for 3 months is another three hundred fifty K or so. Hell, it would cost almost that much in severance and benefits extension, so it's not worth it. Keep the people aboard and find something useful for them to do. Or at least get some productivity improvement by putting the best ones to work and letting the rest do the make-work. The last thing we need is a bloody layoff with union negotiations coming up in December. So that's the answer: yes on the priorities, no on the layoff. Any disagreement?" Tell's answer had the tone of finality, but he would always listen to a counter-argument if offered.

"Aye, aye, sir," Hamilton saluted.

Decker took advantage of the remaining time to bring Tell and Hamilton up to date on the program replanning effort, especially the fact that he had taken it upon himself to shorten the satellite software schedule by two weeks with seventy five percent confidence at best. He pointed out that the meetings with the Navy team starting tomorrow were designed to get them on board with the new program plan. If they had good suggestions, they would be incorporated. He indicated that satellite launch rescheduling was included in the plan and was essential to meeting the end date, and he inquired if Tell had discussed satellite rescheduling with Admiral Sullivan.

"Yes, I did, and he doesn't like it. But I told him that he had two choices: reschedule the damn satellite launches or retire early before the shit hits the fan. He chose the former course of action. They'll give you a hard time about it during your meetings this week, but you'll get your new satellite schedule, Jack. Meantime Rusty and I have our work cut out for us laying pipe for this in Washington."

Tell's uncanny ability to sway those with an opposing point of view – often by direct if unpleasant statement of the obvious – had been relied upon by Decker, and he was not disappointed. The Admiral's support of Decker's initiative had not been a sure thing, so Arnie Tell had come through, as Decker knew he would.

Decker excused himself with the comment, "Thanks, Arnie. Much appreciated. I'll leave you guys to quibble over unimportant non-SEANET stuff. See you at four, Arnie?"

"Right, except I'll be on time," was Tell's get-even closing.

He arrived at Decker's office, in fact, two minutes early, allowing a small margin of error. He was greeted by Erik Hansen, who recognized him at once from photographs in the periodical Tellonics Communicator and from having delivered mail to his office when he worked at Tellonics headquarters in Burbank. Tell took the time to introduce himself and to inquire with genuine interest about Hansen's present circumstances, pointing out that he remembered him from several years ago at Burbank. Tell had a remarkable memory for faces and names, a powerful tool for a CEO. "Pleasure to meet you, Erik. And do me a favor, will you? Tell that boss of yours that I was on time. He won't believe me."

"Of course," Hansen replied with a broad grin.

"Nice looking secretary, Jack," Tell said as he entered Decker's office and closed the door behind him. "When are you having the operation? Or do you need one?"

Decker, usually quick with a barb of his own, just shook his head. "I knew you'd have some smart-ass remark about Erik. Fact is, Arnie, he's the best assistant I've ever had. I'm seriously considering keeping him on permanently if I can get away with it. I may need your help to bend the HR rules a little."

"Me? Bend rules? Never!" was Tell's shocked, if feigned, reply.

Tell inquired about Decker's personal arrangements and was pleased to hear that he had signed a lease on the townhouse. Decker also told him about his dual-office arrangement and let him know that he had received a very warm reception at DSD. Then he asked about the secret.

"The Senior Senator from the Commonwealth of Massachusetts is out to get our collective asses – especially Rusty Sullivan's and mine. Flannery was really pissed when we beat out New England Electronics, but not nearly so much as he was when they lost the protest. He's done a number of unscrupulous things, planted informants and stuff like that. And he has a staffer, Blake Blaisedale, a bulldog of a former prosecutor, keeping track of the program full time at taxpayer expense. So far we've managed to perform well enough that Flannery can't make a stink about it, but when they get wind of our cost and schedule problems and the need to replan the program there's going to be a helluvan uproar in the halls of congress. So stand by for a ram, as they say in the Navy.

"When it hits the press, and it will, tell your people to stay out of it. Plead ignorance and refer all inquiries to the PR office in Burbank. We'll be ready with a press release if it becomes national news. So just keep running the program. Rusty and I will take care of the politics. Unfortunately, the timing ain't all that great what with the boardroom problems I've got. But not to worry, we can handle it." Tell spoke confidently.

"Let me know if I can help with the press release, or anything else for that matter," Decker offered. "We've got a pretty defensible position, actually, as high risk programs go, but it takes a bit of explaining to make the case to the press and the public. We will definitely be the underdog, and that could work to our advantage from a PR point of view."

"OK. I'll have Jean Mosley from PR come down and talk to you. She can be trusted. Don't want any of this on the telephone or the computer system. Any documents concerning this should be hand-written – no copies, so be cool." Tell was concerned about leaks to the press before Flannery went public. He wanted to be ready with a crisp defense of the program to be released at just the right moment, not before. Also, it had to be coordinated with the Navy's release in both content and timing.

He concluded the topic with, "At some point we'll probably have to get the lawyers involved, but I'll keep them out of it as long as I can. In my opinion this is not a legal matter, it's a business judgment matter. We haven't done anything dishonest -- stupid maybe, but not illegal. Let's keep it that way.

"Now what was it you had to tell me?" Tell wanted to know.

"It might help you with your boardroom problem, maybe not," said Decker. "It has to do with Commware and your nemesis Mason Crenshaw. They have been uncooperative as a subcontractor, and our folks have come to suspect that they have something to hide. Despite the fact that you can't stand it, they can't hide their CSSR data. They're running two months late and are about fifteen percent over budget. The schedule problem is due to technical issues, but based on data I've seen, I'm convinced that they intentionally low-balled their cost proposal by almost exactly the amount of the overrun. In other words, I suspect that they put together a sound cost estimate as they are capable of doing, and then their management arbitrarily reduced it by fifteen percent across the board. Why would they do that, Arnie?"

"Well, I'll be damned," Tell replied. "I'll bet that sonovabitch did it on purpose so that he could make the program – and yours truly – look bad to the stockholders. I wouldn't put it past him.

"Look, Jack," he continued. "If you can, I need you to build that case for me and document it. You're right that it could help me on the corporate side, and if we can prove it true, it could help us explain the need to replan the program. I also wonder if there's some connection to Flannery. Rats run in packs, you know."

Decker added that Commware would be in for a standup program review next week and that he might be able to subtly get more information from their program manager if he was an honest guy.

Tell, unable to resist the opportunity for a friendly jab at Decker, interjected, "Whoever heard of an honest program manager?"

"Well, maybe he's an exception. Even so, we'll hold their feet to the fire and find out as much as we can. We have a contractual right to see complete details of their cost data, and their proposal is very detailed, so we can make a comparison. The one thing I don't have yet is their past performance history on software development to demonstrate that they cut their estimate below their standard bidding estimates. But since they've done other cost-reimbursement subcontracts for Tellonics, we have their other proposals to use for comparison. If my suspicions are correct, we can make the case. You'll have it in a couple of days.

"And oh, by the way, I've been calling all the CEOs of delinquent subcontractors to try to get schedule or cost improvements from them. The only one I haven't spoken with yet is Crenshaw. Any guidance?" Decker needed Tell's input before placing the call because of Corporate concerns beyond SEANET.

Tell thought for a moment, and responded, "Yeah. Don't call him. I'll be talking to him myself in a few days after I get your information and will deliver your message ... and mine."

The elephants had spoken.

CHAPTER SIXTEEN
THE NAVY

Jose Alvarez was born in El Paso, Texas, the son of illegal Guatemalan immigrants Juan Alvarez and his wife Maria. His parents took advantage of the 1980's amnesty, obtained green cards, and were naturalized as U.S. citizens in 1989. They were determined that their children, U.S. citizens by birth, would obtain the best possible education, and despite their humble beginnings they were not disappointed. Of the seven, four were college graduates -- the eldest, Jose, from the United States Naval Academy. The occasion of Jose's appointment to the academy by Congressman Manuel Estaban was one of the proudest moments of Juan and Maria's hard-working lives, second only to that of his graduation in the top quarter of the class of 1990.

Since graduation Alvarez had occupied four Navy billets for an average posting of just under three years, including an aggregate of 58 months, nearly five years out of eight, at sea. As an unmarried officer the at-sea assignments were accepted without complaint, and he had attained the rank of Lieutenant Commander.

His most recent assignment was a prestigious posting as Weapons Officer aboard the AEGIS destroyer, USS Port Royal. This assignment by chance brought him in frequent contact with the crew of the nuclear aircraft carrier, USS Jimmy Carter, of which Rear Admiral Russell Sullivan was Captain.

Every U.S. aircraft carrier has at least one AEGIS destroyer as an escort when sailing in harm's way. The sophisticated phased-array radar, AN/SPY-1, is the heart of the AEGIS integrated combat system, serving as its all-weather eyes searching hundreds of miles in three dimensions and capable of detecting and tracking hundreds of air or surface targets simultaneously with great precision. The weapons aboard that ship include surface-to-air missiles, cruise missiles, rocket-launched anti-submarine torpedoes, and a Phalanx Gatlin gun -- the defense of last resort against surface-skimming anti-ship missiles that might penetrate the outer defenses. She also carried traditional weapons such as a five-inch gun, tube-launched torpedoes, and depth charges. In addition to the AN/SPY-1, she boasted sensors including two-dimensional long-range search radar, navigation radar, the Phalanx radar, the 5-inch gun-laying radar, and two sonar systems, one towed in a special capsule trailing the ship. Port Royal could also accommodate a helicopter for transportation, surveillance and rescue missions. Her communications equipment covered the electromagnetic spectrum from very low frequency to ultra high frequency – including even satellite TV for the crew's entertainment. Port Royal's Combat Information Center displayed contacts reported not only by its own sensors, but also those from other vessels, aircraft, and satellites with which it was in communication. The limitations of the latter pointed to one of many needs for a more capable intra-Navy communications system, which came to be known ultimately as SEANET. Directing the operation and maintenance of this complex, fully-integrated weapon system was an awesome responsibility for a young Lieutenant Commander, but Port Royal had the best operational availability record of any AEGIS ship in the fleet.

In addition to his duties as Captain of Jimmy Carter, Rusty Sullivan was Commodore of the entire carrier task force including Port Royal, and the performance of Jose Alvarez did not go unnoticed by the Admiral. A senior Rear Admiral, Sullivan was rotated into his next assignment in January of 1998. As Program Manager of SEANET, a Major Weapon System development program, he reported directly to ASNR&D, the Assistant Secretary of the Navy for Research and Development, bypassing the usual military chain of command and just two levels removed from the Commander In Chief. One of his first actions in this high-profile post was to request the assignment of Lieutenant Commander Jose Alvarez as his aide-de-camp, and Alvarez reported aboard in early February. Sullivan had a dual purpose in that appointment. The administrative and ceremonial assistance of an aide were all very necessary, but what he needed badly was the expert advice of a trusted engineer with at-sea experience and no competing organizational attachments. In Alvarez he got both a personable aide and an unencumbered technical advisor.

The first major decision required of Sullivan was the selection of the SEANET prime contractor. Having studied and evaluated the proposals of the competitors, the Source Selection Board, composed of Navy Department experts and consultants in every relevant discipline, made its recommendation to the Source Selection Authority, Rear Admiral Sullivan. After considerable consultation – especially with Alvarez – Sullivan made his decision, and the billion-dollar contract was awarded to Defense Systems Division of Tellonics as recommended by the Source Selection Board. Two and a half years later the program was in trouble, but Tellonics had appointed a capable new Program Director, Jack Decker, whose revised program plans were ready for review by the Navy. Once again, Jose Alvarez was called to duty.

Alvarez assembled a team of twelve to conduct the review at Defense System Division's Buena Park Campus. As his second in command he selected Earl Riddle, Technical Director of the Navy's SEANET Program Office and a member of Sullivan's staff. Riddle was a career civil servant, grade GS 15, technically proficient although somewhat lacking in administrative skill. Over his twenty-three years of service with the Navy, Riddle had come to view contractors as a necessary evil, but evil nonetheless, profit-motivated and not altogether believable. He had never caught a contractor representative in an actual lie, but he believed that they typically shaded the truth or hid problems to their selfish benefit. Although outwardly friendly with contractor personnel, in Riddle's opinion contractors must be viewed under a microscope to get a valid picture of contract status. The other ten members of the team were subject matter experts in various relevant disciplines from design to logistics support. Major Randolph Garner, an Air Force representative from Vandenberg Air Force Base, was included to review the satellite launch plans. Riddle would rely on them to wield their inquisitive microscopes during DSD's presentation of the revised program plan.

Government employees paid their travel expenses from a fixed daily expense allowance and tended to stay at acceptable but less expensive hotels to fit within its limitations. The per diem allowance had to cover meals, lodging, and all personal expenses other than transportation. It was paid regardless of actual expenses above or below the set amount established for the travel location, one hundred twenty five dollars in the case of the Orange County, California, region. By watching their expenditures, they could usually break even or better. Contractors, on the other hand, were typically reimbursed for their actual expenses so long as they were reasonable with the understanding that luxurious accommodations or gourmet meals or alcohol were not considered reasonable. Having traveled from all parts of the country, the Navy team assembled by prearrangement in the lobby of the Howard Johnson Hotel in Buena Park at eighteen hundred hours, on Monday, August twenty eighth. They were met there by Alvarez and Riddle, who had traveled together and worked out a strategy for the review en route.

The review was expected to take three full days, but they would stay longer if necessary. Tentative travel plans could be made to return home on Friday. The team would meet for a government-only caucus each morning "...from oh-seven-thirty 'till oh-eight-hundred hours..." to set priorities for the day in a secure conference room assigned for their full time use by DSD.

A similar government-only caucus would be held at the end of each day to review findings, exact time and duration to be determined by circumstance. General instructions were to dig into the details of the contractor's plans to determine if there were any flaws that needed to be addressed and to offer constructive suggestions. This was to be a critique, not a confrontation, since it was in the Navy' best interest to have the contractor succeed, not fail. A debriefing would be held on the last day, and the review team's preliminary findings would be presented to the contractor for comment. Written reports should be sent to the team leader by encrypted email, eyes only, not later than Tuesday morning, one week from tomorrow. Based on those reports and their own observations, Alvarez and Riddle would brief the Admiral next Tuesday afternoon.

The Navy team members arrived more or less simultaneously at the DSD guard shack around seven fifteen Tuesday morning in four rental cars. They were expected, and after showing ID to the guards, were directed to the visitors lot at DSD headquarters, the glass and brass building, where parking spaces near the entrance had been reserved for them and marked "SEANET REVIEW." As they entered the lobby of DSD headquarters, Ken Martin and Milt Karinski met them and presented them their "no escort required" visitor badges by prearrangement with security, their secret security clearances having been faxed to DSD Security by each of their own security organizations. They were shown to the secure conference room designated for their use, where a guard, assigned for the duration of the review, logged in each of them after examination of their visitor badges and picture identifications. The main conference room, also guarded for authorized entry only, where the contractor briefings and discussions were to be held, was a few doors down the hallway.

Hot coffee and tea, chilled fruit juices, and a variety of pastries were laid out on a side table near the entrance of the main conference room, and as the attendees entered and introduced themselves to others, they helped themselves to a light breakfast – the second for some of them. This brief get-acquainted period was an essential part of the review, since these people, Navy and contractor alike, would be locked together for the next three days of intense mental endeavor and needed to get to know each other to enhance open communication.

Ken Martin asked everyone to take a seat at about ten minutes after eight. The twelve Navy team members were invited to sit at the long conference table, and Jack Decker, Ken Martin, and Milt Karinski occupied three of the four remaining chairs. DSD's other SEANET team members took seats around the periphery of the room. Then Martin began the meeting with his opening remarks.

"Good morning everyone, and welcome to sunny southern California, even though you may not have much time to enjoy it." Martin then took a minute to outline the logistics arrangements for the meetings: a light breakfast would be available here each morning as it was today, and a hot working lunch would be served buffet style in that room each day from noon until one in order to conserve the review team's valuable time. These amenities were believed to be well within government guidelines for acceptable contractor gratuities, but if any of the visitors felt obliged to pay, the charge was two dollars for breakfast and six for lunch. A container was available on the table for that purpose. Martin did not expect them to pay but was required by strict interpretation of federal regulations to offer them the opportunity to do so. At the end of each day, contractor representatives would drop by the Howard Johnson "Olde English Pub" for refreshments if any of the Navy team would like to join them for a social hour – this definitely not financed by the contractor.

Jack Decker interrupted. "As you know, it's against the law for me to buy you a beer, but it is perfectly legal for you to buy me unlimited Johnny Walker Blacks on the rocks." The tension was broken for the moment as everyone present joined in laughter.

Martin continued with, "So that's the general outline of the arrangements. We'll begin our overview of the revised program plan in a few minutes, but first I'd like to invite Commander Alvarez to make any opening remarks he may have."

Alvarez thanked Martin and Decker for their hospitality, and then said, "I'd like to defer the Navy's opening remarks to our boss. I believe that Mister Decker has made arrangements for a phone call to the Admiral." Decker dialed the number, and spoke briefly with Jane Wilson who put Sullivan, standing by for the call, on the line as Decker energized the speakerphone.

"Good morning everybody," Sullivan began, "this is Rusty Sullivan. Jack Decker has graciously afforded me this opportunity to speak with you all in the hope that I'll say something favorable about him and Tellonics." Another short round of laughter interrupted the Admiral. "Well, I would really like to do that and I will, but only after we have agreed on a plan for continuing the SEANET program with a much-improved probability of success. Jack has given me a brief overview of his recommendations, and they sound reasonable to me. But what do I know? I'm just an old fighter pilot. It's up to you guys and gals, Government and contractor alike, to put together the best damned plan you know how. It's not to be DSD's plan, or the Navy's plan, but our plan – our collective plan. Once that's done, and we can see some light at the end of this dark tunnel we find ourselves in, I'll have some nice words for you all – even Decker." As the audience responded with a brief chuckle, he continued, "It's an overworked quotation, I must admit, but I can't

184

think of a better one: 'let us all hang together, or we shall certainly hang separately.'

"One last thing. I do not want any leaks. The fact that we are replanning this program is very sensitive and will remain so until it is completed. At the appropriate time and by appropriate means, we will go public, but not before. When we go public, we will do so with a realistic plan that is defensible against our critics and competitors for funding.

"With that, I'll sign off. But I'm available if you need me. I have nothing more important to do than to offer you my full support for what you are about to do."

"Thanks for your comments, Admiral Sullivan," Decker directed his voice toward the speakerphone. "We'll be in touch to let you know how things are going. And we'll take you up on that offer if we need your help."

After hanging up, Decker directed his remarks to the group assembled, his message intended as much for the DSD personnel as for the visitors. "I don't know if our visitors noticed, but they have "no escort required" visitor badges. That's significant, because we don't usually do that for visitors. The message is, we have no secrets regarding SEANET. You are free to wander about these premises as you choose except for certain restricted areas outside your need to know. This is an open kimono meeting as far as we are concerned, and we'll share with you every bit of information we have that will help you in your mission this week. There is only one possible exception, and that is proprietary information. There are a few trade secrets, technologies, processes and inventions that we don't share with our competitors. If they are related to SEANET, you can have access to them as Government representatives, but with the understanding that they will be held closely by you as contractor proprietary information. It's up to us to designate in writing or by appropriate markings when data are proprietary. The rule is: if we don't say it's proprietary, it ain't proprietary.

"I'll be in and out of these meetings, joining you from time to time in your various discussions. But like Admiral Sullivan, I have nothing more important to do this week than to assist you in every way I can. So Ken, let's get started." With that, Decker turned the meeting over to Ken Martin.

The visual aids for the briefings were full-color computer graphics images projected onto a large-screen display. A remote control device held by the presenter allowed selection and sequencing of the briefing charts. For many years transparencies displayed by overhead projector, viewgraphs as they are called, had been the standard for professional presentations, but DSD like many others had moved into the computer graphics world. An entire presentation consisting of hundreds of equivalent full color viewgraphs could be recorded on a single ZIP disk and transported to the presentation location. For this meeting, hard copies of the computer graphics had been printed for the reviewers, bound in inch-thick books to make the review process as easy as possible for the Navy team.

Program Overview briefings were presented by Martin and Karinski and occupied the rest of the morning. They summarized the work that had been done by DSD up to this point in the program and what lay ahead. The last and most important topic was a summary of the revised program plan that had been worked out by DSD the week before. Particular emphasis was placed on changes, whether cost, schedule, or technical, that were being proposed by DSD. After lunch, more detailed briefings were presented by several of the DSD Department Leads in much the same fashion as had been done for Jack Decker's review the previous week. In those presentations the specific changes to the program plan and the rationale for them was given. The potential solutions to technical problems were detailed, and schedule status reports were provided for items on or near the critical path. High interest items, as expected, were the Commware software timing problem, the satellite syncodec heating problem, and the microwave board manufacturing problems.

The Navy representatives had a number of serious questions concerning those difficulties, and it was decided that rather than tie up the entire group, splinter sessions, that is to say separate meetings attended by those responsible or interested, would be held the next day for each of them. The most prolonged general discussion, however, concerned the proposed system integration and test schedule presented by Tony Grazio. It was in this presentation that the new satellite launch schedule was proposed. Grazio explained the necessity of the delay in launching and showed that the overall test schedule could be rearranged in order to meet the required end date. This, too, prompted the need for a splinter group to better understand the problem and its solution, Major Garner being especially interested. When the day's briefings were completed, it was almost five o'clock, and the Navy retired to its secure conference room for its end-of-day debriefing.

Jose Alverez convened the Navy caucus, gave his own impression of the day's events, which were generally but not entirely favorable, and asked for comments by the others. Notably, Earl Riddle commented that, contrary to his usual experience with contractors, he had been impressed by the candor of the presenters and hoped that DSD would continue to be as open as Decker had promised. He would withhold final judgment until the review was completed. To conclude this first day's business, Alverez requested that the members be prepared to outline their plans for the Wednesday meetings when they met in the morning.

Jack Decker entered the Olde English Pub just after six o'clock. He was pleased to see Martin, Karinski, and many of the DSD SEANET Department Leads present, as were most of the Government representatives, including Alvarez, Riddle, and Garner, the three whose opinions would carry the most weight.

Waving to the group, he approached the bar to order a drink. Before he could speak the bartender smiled and placed before him six pre-prepared tumblers filled with Johnny Walker Black Label Scotch Whiskey over ice. "Compliments of the United States Navy, Mister Decker," the bartender said -- loud enough for everyone in the bar to hear.

There was a grand round of applause as Decker lifted the first (and last, he resolved) glass to his lips in salute. "Cheers," he toasted, with an embarrassed grin.

The social hour prompted lively conversation among the attendees. On both sides of the contractual wall, enlightened management had found that business could be conducted best by people who understood each other as individuals; however, business per se was not to be discussed at such events. Decker made it a point to have a personal conversation with each member of the Navy team, including ten minutes in a three-way conversation with Alverez and Riddle in which they discussed how they had become slaves to their personal computers. He noted that Alverez, likewise, spoke with each member of the DSD staff.

As arranged by Decker, the DSD team members departed one by one such that by seven o'clock, having consumed less than the legal limit of alcohol for safe driving, all had left. Decker had insisted that the Navy team be protected against the appearance of excessive contractor influence leading to charges of a conflict of interest by those who did not share their enlightened view. Decker said goodnight to all and was the last to leave, insuring that no DSD stragglers stayed behind.

The second day of the review was devoted mostly to separate meetings of the several splinter groups composed of subject-matter experts in their fields of interest. Alvarez and Riddle for the Navy, and Decker, Martin and Karinski for the contractor, rotated among the meetings separately to get the gist of the discussions, joining in and offering guidance as they saw the need. After lunch the same five had a one-hour meeting of their own apart from the others, organized purposely by Decker and held in his VP office. He wanted to know what issues were of most concern to the Navy. Alverez was hopeful that the splinter groups could conclude on their own the plausibility of DSD's proposed solutions to current problems in cost, performance, or schedule, but he was concerned about the political difficulty of rescheduling the satellite launches. Riddle indicated that he had two principal concerns: the need for relaxing the specifications and the proposed system integration and test process.

Riddle had reservations about the plan to defer certain subsystem tests to be conducted concurrently with full system tests – the leapfrog strategy, as Decker had called it. Decker agreed with him that this was a higher risk approach but affirmed that it would only be done when the risk was believed to be acceptable. He gave a number of examples from his personal experience, pointing out occasions when it had been an acceptable risk as well as when it had not. Riddle agreed that in order to compress the schedule by as much as two months some risks had to be taken, but he wanted to be sure that the risks were well understood and manageable. The others joined in with their views on the subject, and at the end of the hour, Riddle appeared to be convinced that this was an acceptable way to proceed provided the Navy had a say in which tests were to be "leapfrogged." He would consult with other team members to establish guidelines for implementing the proposed strategy.

A tour of the Defense System Division's SEANET facilities had been arranged for mid-afternoon, and at three o'clock the entire group reassembled in the main conference room. There the visitors were divided into three groups, each independently escorted by DSD Department Leads acting as tour guides. In groups of just four each, the Navy reviewers would be able to get a clearer picture of program status and to inquire about what they saw more directly. At each station of the tour, knowledgeable briefers were standing by to demonstrate the specific hardware and software being assembled and tested. Stations in which the visitors showed considerable interest were the software integration facility, where a complete system could be simulated using much of the actual system hardware and software, and the Land Based Test Site, in which the final product would soon be integrated and tested prior to installation in Navy platforms.

The highlight of the tour was the large clean room in which the satellites were being assembled. It was obvious that the first satellite had been fully assembled, its access panels having been removed for internal visibility, and was ready for the installation of final software and retesting of the syncodec cooling system once the specification issues were resolved. The other two satellites were also well into the painstaking assembly process. It was an impressive exhibition of state-of-the-art technology under development.

The second day of meetings ended with the afternoon Navy Caucus. Unlike the first day, however, there was much to report by the splinter group attendees. They were convinced that they were getting the complete picture, that there were no hidden agendas, and that they could rely on the contractor's good intentions. There were a number of areas, however, in which they would propose alternatives to the contractor's plans. Discussed openly, many of them had been accepted by the contractor. Except for two, the splinter groups had concluded their independent meetings and were ready to reassemble in the main conference room the next day. The satellite software group and the system integration and test groups needed another hour or so to conclude their discussions and could join the general meeting at ten or eleven o'clock tomorrow, Thursday, morning.

The Olde English Pub greeted essentially the same group that had assembled there on Tuesday evening. On this occasion, however, the larger group dissolved naturally into several smaller ones, each populated with splinter group attendees who had now become better acquainted. Decker and Alverez paired off for a time and then were joined by Riddle and Garner. It appeared to Decker that this was not an accidental encounter, so he listened carefully to what was being said, saying little himself.

"I do think we'll be able to complete our review tomorrow, Jack." Alverez was now addressing Decker on a first name basis. "If our subject-matter experts can complete their reviews in the morning, we should be able to debrief you on our initial findings by two o'clock or so. Then we'll be out of your hair for awhile."

Major Garner added, "I wonder if you could make a trip to Vandenberg next week. I may need a little help with my 'Admiral', who happens to be a pretty sharp Air Force Lieutenant General."

"Of course." Decker said. "Just let me know when and where. A little background information on your 'Admiral' would be helpful, too, if you can provide it. Of course, I'd need my Admiral's approval – politics, you know. In fact, maybe he should come along."

"Now that's a fantastic idea!" Garner responded. "Why didn't I think of that?"

Riddle interjected, "That's what the big bucks are for, Randy."

Alverez thought it a good idea, too, "I'll talk to the Admiral tomorrow and see if he's available. You could mention it in your daily telecon, too, Jack. It's also possible they could meet in Washington depending on their schedules."

That was the full extent of business discussions during the social hours – just setting up another meeting.

As promised, Alverez and the Navy review team concluded their independent meetings and were ready to present their preliminary findings to Decker and the DSD SEANET team at two o'clock Thursday afternoon. "To summarize," he began, "we think you folks have done a pretty good job of restructuring the program. I think it is fair to say that we are in general agreement with your overall strategy. We also have some serious concerns here and there, so we will have some specific recommendations to make to Admiral Sullivan that will probably result in some contract changes. As you pointed out back in Washington two weeks ago, Jack, we need to keep the program and the contract in sync. What I'd like to do this afternoon, then, is to have some of our people give you their preliminary findings so that you can comment on them before we take them to the Admiral. Hopefully we won't have any showstoppers. Let's start with Earl Riddle."

Riddle indicated that he would recommend several of the specification changes that DSD had requested but would not change any of the system's essential performance parameters. He saw potential problems lying in the path of success not included in DSD's integration, test, and evaluation replan, but he had brought them to the attention of the responsible people and believed that they could be resolved to the satisfaction of both parties. On the subject of "leapfrogging," he was now a convert, but wanted to retain approval authority for each "leap," as he put it.

Major Garner delivered his debriefing next. As representative of the Pacific Missile Range responsible for the satellite launches, he indicated that he understood clearly the status of the satellite production schedule. He would personally recommend acceptance of the contractor's revised satellite launch schedule, but knew in his heart that it would require more than just his recommendation to secure approval. He would need the help of both the Navy and contractor SEANET Program Managers to convince his superiors of its necessity. Until then it remained an open issue.

Decker thanked Garner for his support and indicated that he was ready to help with selling the need for a delay in the satellite launches. He resisted the urge, however, to point out that even if the schedule were not changed, Tellonics would never launch a satellite known to be deficient – even if the Navy would allow it, which it would not. That message would be better delivered to a higher-ranking audience.

Three other members of the Navy review team delivered their debriefings. Two were fully supportive of DSD's proposed replan. The third, delivered by GS-13 Scott York, subject matter expert on microwave circuitry, pointed out that neither he and his colleagues nor the DSD "gray beards" had so far identified the cause of the microwave circuit card production problems. He would have expected a "glitch" or two in such intricate, precision circuitry, but to have all four exhibit very similar problems was strange, indeed, and beyond his experience. Obviously, this problem had to be resolved before it found its way off of the critical path, and he would continue to consult in its solution.

Linda Carpenter, GS-12 subject matter expert on Ada software, expressed her concerns about the satellite software and its timing problems. She was skeptical about the Ada to C++ translator, but she was glad to hear that DSD had two potential solutions being worked in parallel. Milt Karinski commented for the benefit of others in the room that initial test of the X500 microprocessors with the Ada software loaded had been promising, but more testing was required.

To conclude the review, Alverez thanked Decker and the DSD representatives for their thorough and forthright presentations. He would recommend to the Admiral a revised program plan not very different from that proposed by DSD.

In his closing remarks, Decker indicated his appreciation for the orderly and very professional way in which the Navy team had conducted its review. He looked forward to working with them to make SEANET a system of which the Navy would be proud. With that, he suggested that they reconvene in the Olde English Pub for a bon voyage meeting.

On Friday, the Navy review team members returned to their home stations. On the return flight to Dulles International Airport, Alverez and Riddle agreed that it had been a very busy but very productive week, and they began the preparation of their briefing to Admiral Sullivan.

Before leaving the plant Friday night, Decker stopped into the VP office to clean up the paperwork that had accumulated due to his preoccupation with the Navy review. Erik Hansen had asked permission to leave a few minutes early, and as he departed, almost as an afterthought, Hansen said, "One last thing, Mr. Decker. The IT folks will be in over the weekend to upgrade your computer and install some additional software. They asked that you leave your office unlocked if possible."

Decker said, "That'll be fine. I don't usually lock the office door. Anything sensitive I keep in this battleship of a desk – it's got a superb locking mechanism – and besides, I won't be using the office over the weekend anyway.

"Have a great weekend, Erik."

CHAPTER SEVENTEEN
LONG WEEKEND

Jack Decker welcomed the long Labor Day weekend. Although he had enjoyed it for the most part, the last two weeks of long days had been packed with intense activity. He needed to relax; he also looked forward to seeing Susan Anders.

The first order of business was settling into the town house. On Tuesday the telephone, utilities, and television/computer cable had been placed in service, so on Wednesday morning he had checked out of the Marriott and transferred his bags to the town house before reporting to work. After signing the lease last Saturday, he had called his son in Philadelphia and asked him to ship the six boxes that he had packed previously at the Bryn Mawr house. The boxes contained additional clothing and personal items that he would need, and they had all arrived on Friday, accepted in his absence by pre-arrangement with a new neighbor. Although completely furnished, including ample linens and kitchen ware, there were some additional things needed. The pantry would have to be stocked with the usual staples, and groceries for the week would have to be brought in.

Although he had become accustomed to living alone with minimal domestic help after Marty's death eleven years ago, he was glad that Susan had volunteered to help him "set up housekeeping" and would arrive about ten o'clock Saturday morning for that purpose. They planned to spend the weekend together at the townhouse, the beach house, or whatever came to mind, possibly including a drive up the Pacific Coast Highway to Malibu or beyond on Sunday.

+++++++++++++++++++++

Draftsman Patrick Riley had also made plans for the weekend. Realizing that few others would be working over the long Labor Day weekend, he had decided it would be an ideal time to come in to DSD for his special mission. Few would be working overtime over this last long weekend of the summer. With his secret security badge and parking permit, he had no difficulty entering the DSD grounds. He was required, however, to sign an after hours log at the front guard shack as were all those who entered or left the premises at times other than normal working hours. Since there were no guards at the individual buildings during these off hours, the gate guards would perform the random inspections of briefcases, purses or packages for classified documents, cameras, recorders, and other devices prohibited for security reasons. Occasionally vehicles would be searched on a random basis. Riley was prepared for this eventuality, having been so processed previously, but he passed through this time with no inspection. "The luck of the Irish," he thought.

The parking lot was almost empty, and since parking area assignments were not enforced on weekends or holidays, he parked the silver Ford Explorer SUV in the visitors' lot adjacent to the DSD brass and glass headquarters building. With no receptionist or security guard on duty at the entrance on weekends, Riley proceeded directly to the executive office area without being observed. He found the nameplate, Jack Decker, at the entrance to Decker's executive suite without difficulty. There was no one working in the area. He put on a pair of surgical gloves to avoid leaving fingerprints and entered Decker's unlocked office unnoticed. He carried with him the alternate briefcase that had been carefully hidden in the SUV to avoid detection in the event of a vehicle inspection. The briefcase contained a typical assortment of hand tools that might be carried by a computer technician, and in a false bottom compartment it also contained the homemade bomb and detonation apparatus.

It was similar to a pipe bomb. He had informed himself on their construction from articles in the radical publications to

196

which he subscribed and even on the Internet, and he had constructed a compact and lighter weight variant of that dark technology. The explosive was contained in a flat metal electrical box rather than a length of pipe. The detonator was a blasting cap inside the box that could be activated with an electric current, and two wires from the detonator circuit extended through insulators at one end of the box for that purpose. Riley had devised a clever but simple method for providing the necessary current to the detonator. The outer insulation of a three-wire extension cord had been slit to expose a short length of the three independently insulated wires inside. One hundred turns of thin magnet wire had been laboriously wrapped around the black "hot" insulated wire to make a transformer. The outer insulation had then been repositioned and held in place with transparent tape, with a few feet of the fine magnet wire extending from the now unnoticeable wound. An inch of the lacquer insulation was carefully stripped with a knife from the ends of the magnet wire exposing a hair's width of bright copper, and each of the wires was wrapped around one of the wire ends protruding from the metal box. The doctored extension cord would be placed in series with the power cord of the computer equipment so that when the computer was turned on and a few Amperes of alternating current passed through the black wire of the extension cord, a small alternating current would be induced in the wrapped turns of magnet wire. A diode rectifier and electrolytic capacitor inside the box would provide sufficient direct current voltage to activate the blasting cap.

Riley had constructed several prototypes of the detonation device and tested them in a remote area using a small, gasoline-operated emergency generator. After optimizing the number of magnet wire turns and the size of the capacitor, the blasting caps could be detonated without fail with as little as one Ampere of current in the extension cord.

He located Decker's computer on the desk, and found that it and its peripheral accessories were plugged into a single power bar next to the printer/scanner/fax device. The switch on the power bar was in the OFF position, so it appeared that Decker routinely turned his computer equipment on and off with that single power bar switch. It was an ideal arrangement for Riley's detonator. He installed the modified extension cord between the power bar cord and the floor outlet under the desk and connected the magnet wire transformer leads to the wire terminals extending from the flat box. The box was attached with duct tape to the underside of the desk just behind the center drawer so that it was hidden from view, and the magnet wire leads were so fine as to be virtually invisible. The presence of the short extension cord under the desk was unlikely to be noticed.

The entire operation had taken less than five minutes. Riley left Decker's office and returned unobserved to the SUV, and then he drove to the parking lot nearest his regular work place. He intended to do some computer work there for a few hours in order to be able to account for his presence at DSD on that particular Saturday. He made it a point to be seen hard at work by two others who were also working that day and would recognize him.

Riley was certain that Jack Decker would be killed or at least severely injured when he turned on his computer on Tuesday morning. He was also convinced that he could not be connected to the event. This was the second of his sabotage actions against SEANET, the first having already taken effect undetected. The third was a phone call that would be made early Tuesday morning.

+++++++++++++++++

Susan arrived just after ten o'clock Saturday morning

and parked, as instructed, in the underground parking area below the complex of townhouses. Each of the residents had a magnetic card that could be inserted in a card reader to open the gate. There was also an intercom that could be connected to the resident of choice who could then open the gate from inside the townhouse – the method used by Susan Anders on this occasion. As a final measure, there was a keypad that would accept a secret code known only to the residents, one hoped, that could be used to open the gate in case one's magnetic card was misplaced or forgotten. A similar arrangement permitted exit from the underground garage. Also well inside the gate, beyond arm's reach, there was a "panic button" that would open the gate in case emergency exit was required. The children of the residents delighted in activating it to the continuing annoyance of the property manager since it required a manual reset which only he and his staff could apply. It all seemed a bit much to Jack Decker, but this was California.

Susan thought the townhouse an excellent choice. After a guided tour, she took a second independent tour with pencil and paper in hand and noted a few things that might be done to make the place "a bit more cozy." Decker welcomed her decorator's instincts since his own were admittedly deficient, but he did take charge of the kitchen since he had more than a little experience in that domain as a ten-year single parent. He had also decided to upgrade the stereo system, actually to replace it with something much better, since some of his favorite compact discs had arrived in one of the boxes from Bryn Mawr. All told, a full day's shopping was in store for the two of them.

Some of the shopping they did separately, some together. Among the things they did together was the last item of the day, grocery shopping. This was left until last so that food would not spoil in the heat of Jack's automobile on a hot day with Santa Anna winds, as the sometime easterly currents across the desert were called. Suddenly, as Jack pushed a shopping cart down an aisle and Susan loaded it with more than he thought necessary, there was a shout from behind them.

"Jack Decker!" The voice called out. "This is about the last place I expected to run into you." As Decker and Anders turned, he could see that the voice belonged to Steve Murphy, the SEANET Program Control Manager. He was accompanied by an attractive woman who held a six or seven year old boy by the hand.

"Well, hi Steve," Decker said. "I guess you have now discovered that even the front office guys have to eat."

"Even so, it's a nice surprise to see you," was Murphy's sincere response, after which he began the introductions. "Jack, this is my wife Andrea and my son Darien ... Jack Decker, dear, the new SEANET Program Director."

"I'm delighted to meet you Andrea, and what a handsome young man you have there. Hi Darien," Decker spoke to the boy with hand extended. Darien took Decker's hand and shook it vigorously as they all smiled.

"And this is Sue Anders," Decker said without hesitation or obvious embarrassment. He believed it very unlikely that Murphy would know that Susan was the CEO's assistant, and even if he did it was even less likely that he would have contact with the people at Tellonics headquarters. He would assure Susan to that effect when they were alone.

Susan greeted the Murphys with a warm smile, extending her hand to each of them in turn, Andrea first, then Steven. And to Darien she said, "I guess you are just about ready for the new school year. What grade will you be in?"

The boy said shyly, "First grade."

"Well, I'm sure you're going to be a very good student and learn a lot." Susan hoped that by focusing her attention on the boy, she would avoid any questions about being with Jack Decker.

"We won't hold you up, Jack. Really good to see you off campus, and nice to meet you, Sue." Murphy concluded the conversation as they smiled at one-another then parted, Susan gently brushing Darien's cheek with the back of her hand as they passed.

"What a nice family," Susan said to Jack when they were out of earshot. "But I guess we've been found out."

"I wouldn't worry about it Sue. They probably don't have any contact with the Burbank bunch, and by dinnertime they won't even remember your last name. You'll just be remembered as that beautiful lady with Jack Decker named Sue."

"Why, Jack Decker! I do believe you're flirting with me," Susan joshed him with a make-believe southern drawl. "And now that we've been discovered, we may as well finish up here and head for home so we can give them something to gossip about."

"Who's doing the flirting, here?" Jack said as they approached the checkout counter.

****

Sunday morning after he had prepared breakfast for the two of them – "In my house, I do the cooking," Decker had scolded – they drove their separate cars to the beach house in Manhattan Beach. They had decided to spend the late morning on the beach, and then drive up the coast in the afternoon. Depending on the holiday traffic and how the trip progressed, they might stay overnight at some nice place. The plan was to not have a plan.

There was an unexpected surprise waiting for Decker at the beach house. Susan knew that Jack adored uncommon automobiles, had owned several, and enjoyed working on them as a hobby. After returning from the beach, packing a few things for a possible overnight stay, and dressing for the afternoon excursion, Susan said, "Oh, by the way, Jack. Mark agreed that we could use his car for our trip."

"Really? Well that was very nice of him. Just what sort of car is it?" Jack was somewhat curious, expecting that Doctor Anders would drive a late model Mercedes, Cadillac, Lincoln or possibly a Lexus. For a lengthy drive any of them would be slightly preferable to his company-leased Buick, a vice-presidential perquisite, or Susan's Toyota – both nice commuting cars but ordinary. He followed Susan out to the front of the house, she opened the garage door, and there it was.

"Well, I'll be damned," Jack breathed aloud in admiration of the spectacle before him. It was a Jaguar E type open two-seater, at least thirty years old, but showroom new in every respect. It was a canary yellow that glistened so brightly that the reflection of the sun behind him caused him to shield his eyes. With an XK engine developed for the Jaguar D type that dominated the Le Mans grand touring races of the nineteen fifties year after year, the XKE was, in Decker's opinion, the finest sports car ever produced in quantity. Some seventy thousand were built over the years, hardly a month's output for the popular Ford Mustang of those days. Its twelve-cylinder V-12 engine ran so smoothly that it was necessary to observe the tachometer to be certain that the engine was running. Although Jaguar had come close on other models, the type E's famous suspension was never duplicated. With a body designed by aeronautical engineers, it hugged a curve like a child hugged its teddy bear, and this later model's 5.3-liter engine provided sufficient torque to accelerate the machine from zero to sixty miles per hour in just over five seconds. Power steering was standard for the later models like this one, but Decker was pleased to see that the available automatic transmission had been dismissed in favor of a short-throw, five-speed stick shift. As he examined the car with obvious delight, he said, "I'll drive."

"I though you might," Susan said. "Just be careful you don't get a ticket. This car attracts a lot of attention even when it's standing still, and the CHIPs are well-acquainted with it." She referred, with amusement, to the several California Highway Patrol officers who had pulled the car over, lights flashing, just for a closer look.

"I didn't realize your brother was a sports car buff," Decker said.

"He isn't," Susan informed him. "In fact, he drives a comfortable old Chevy that he likes. He got the car by a stroke of good fortune. When he was a young intern at LA General, a guy was brought in from a horrible motorcycle accident. Mark was on duty when they rolled him in half-dead and bleeding internally, and while the administrators were arguing about his lack of insurance coverage, Mark and the nurses saved the guy's life. I've forgotten his name, something Italian, but it turns out that he later became a famous racecar driver. When he died two years ago, Mark got a call from the guy's lawyer informing him that he had inherited the Jag. I guess nice guys do win sometimes after all."

The drive up the coast was a delight in every respect. The weather, the car, the scenery, the company – everything was perfect. California One, the Pacific Coast Highway, or PCH as Californians call it, parallels the coastline much of the length of the state with a few interruptions from time to time due to landslides and earthquakes or to service a slightly inland community. For long stretches it is in sight of the Pacific beaches, at times wandering further inland when necessary, but even then a lesser road can be found near the ocean much of the way. Most of PCH is two lanes wide, but some stretches are four lanes, and a few miles of it rise to the stature of a limited access divided highway, a freeway in the California vernacular. The most spectacular part of PCH is the run from roughly Big Sur to Monterey. There the road is carved out of the high cliffs overlooking the rough Pacific waters as they wash rocky outcrops of every size and shape, each one approached more astonishing than the last.

The territory through which the Jaguar first transported Susan and Jack on this occasion was the California coast in the Los Angeles vicinity. Here the coastline runs almost an East to West course, and the land is flat along the beaches where the mountains have slipped over centuries into the sea. Some not-so-gradual erosion can also be found in the form of rather abrupt cliffs, as in the Santa Monica to Pacific Palisades vicinity, with just enough space between the cliffs and the ocean for PCH and the beach. Even there the sun-lovers manage to find a way to construct their beach houses, many on stilts.

Departing Manhattan Beach at one thirty, they drove along the Vista Del Mar route passing on the seaward side of locales with familiar, some even famous, names: El Segundo, LAX, Playa Del Rey, Marina Del Rey, Santa Monica, the Palisades, Topanga, Las Flores, and Malibu, which they reached at just after four. Ordinarily the trip via the San Diego Freeway could have been accomplished in an hour, but theirs was a leisurely drive with frequent stops to observe the sights and the people. In return, they and the yellow Jaguar were the focus of attention of those being observed. Jack suggested that they drive another hour or so up the coast because he remembered a delightful spot just beyond Point Dume and Zuma Beach where they might decide to dine and stay overnight. Its exact location was uncertain, but he believed he would recognize it. They drove a bit further than he had remembered, beyond Trancas Canyon, and then he spotted the sign on the left. PCH was a few hundred yards inland at this spot, and trees and shrubs obscured the ocean, but the side road led directly to the Seaside Inn. It was exactly as he remembered it.

The Inn was built in the style of the Florida Keys: white clapboard siding, only two stories high, a cedar shingle roof, and screened in porch all around. Fortunately, because of a last-minute cancellation, there was a room available on this long holiday weekend. The room was large and spotless, the décor a rustic nautical style reminiscent of the Captain's quarters on a three-masted clipper ship. A wall of windows with real louvered shutters faced the ocean, and the view was breathtaking, sailboats in abundance as the sun neared the horizon. "You seem to attract exotic sunsets, Sue," was Jack's observation.

At seven o'clock they had a cocktail at the bar while they waited for their table. Susan had her chardonnay and Jack his Scotch whiskey. He saw a single malt, Glenliven, of which he was particularly fond, on display in front of the bar's long mirrored wall, and he ordered it without hesitation – on the rocks, of course. Johnny Walker Black Label, defaulting to Red Label, was always a safe choice in uncharted waters, and others would do, but the single malts were often special, and this one was superb. Susan was offered a sip, which she took sparingly and gracefully but with a look of surprise as the smooth, aged brew warmed her throat. "Wow," she reacted. Now that's booze! If you don't mind, I'll just stick to my fermented grape juice."

Dinner was excellent also, as Jack had expected it to be, and Susan could not say enough about his choice of the Seaside Inn. "I guess this is our first real date, so to speak, although we got things a little out of order. We'll have to come here again, and often." Jack was pleased that Susan was pleased, and as they retired for the evening he complimented the restaurant staff and rewarded them with an unusually large gratuity. He had been known to leave little or no tip at all when the service was poor, so on occasions such as this he was more than generous.

They slept late on Monday morning, Labor Day, ordered a room service breakfast, and departed the Seaside Inn at noon. They decided go up the coast just a bit further, then take the freeways home in order to arrive at the Manhattan Beach house by mid-afternoon. Jack would have another hour's drive beyond that, and they both had to work tomorrow. So they proceeded up PCH to Point Mugu, site of the Navy Air Station that is a key facility of the Pacific Missile Range, then across Pleasant Valley, home of the strawberries, on local roads to pick up the Ventura Freeway at Camarillo, just ten miles inland. The Ventura Freeway, California Highway 101, took them southeastward through Thousand Oaks and Canoga Park, past Reseda, and to the intersection with the San Diego Freeway, Interstate 405, at Sherman Oaks. The San Diego Freeway took them to the cutoff to Manhattan Beach, and they arrived at the beach house just after three o'clock. "Well done, Jack, and not a single ticket. Although you earned one a time or two on the straight-aways."

"A car like this needs some exercise. I was just doing my part," he replied. "Tell Mark that when he's ready to sell, I'm ready to buy."

They relaxed for a while on the patio before Jack left at four o'clock. He was home by five, and poured himself a Johhny- black-rocks to relax. It had been the most enjoyable weekend he could recall, but now there was work to do. The complete Bach Brandenburg Concertos on three CDs were loaded into the new stereo system, he sat at the comfortable sofa, and he opened his briefcase on the coffee table. There were at least two hours of accumulated reading to do, the Navy review having disrupted his routine this past week, and he set about the task.

Susan Anders returned to her Burbank apartment, arriving there at five also, traffic on the freeways having slowed considerably in mid-holiday. Tired, she fell asleep, contentedly, after having turned on a History Channel documentary to pass the time before preparing a dinner snack. She awoke at eleven, decided to forego the snack, and went to bed. She needed to get to the office a little early Tuesday to make some travel arrangements for Arnie Tell.

Patrick Riley's long weekend had been less glorious. After leaving DSD Saturday shortly after noon, he drove out to the desert and spent the weekend in solitude. Returning Monday afternoon, he retired at nine o'clock. He had to get up early Tuesday morning to make an important phone call at eight o'clock east coast time, five o'clock Pacific. Then he wanted to arrive at DSD by seven to be present before Jack Decker sat down at his desk and turned on his computer.

CHAPTER EIGHTEEN
FORCE MAJEUR

There had been quite a stir in the Engineering Department when Jack Decker arrived at DSD as the new SEANET Program Director. Almost from the first hour, rumors circulated that change was in the air. At first there was apprehension that there would be a significant reorganization, but Decker had killed that rumor during his first meeting with the SEANET staff during which he informed them that his strategy was to reorganize the program, not the people. Although they had been instructed to hold close the fact that a major replanning of the program was to be accomplished, and to share that fact only with those who had a need to know, it had been necessary within the Engineering Department to consult numerous people with specialized knowledge. Among them design draftsman Patrick Riley had been taken into the confidence of Lee Chang because of his role in documenting the detailed design of much of the microwave equipment. From Chang's unwitting remarks Riley suspected that the entire program was being replanned. His suspicions were confirmed when he observed that not just Chang but also other SEANET engineering staff were putting in very long hours. Then the Navy review team arrived and stayed for most of a week. From this Riley concluded, correctly, that a major replanning of the program was indeed in the works, most likely occasioned by serious cost, schedule, and/or technical problems, to the latter of which he had been an undetected contributor. This was the stuff of a major whistle blowing.

Riley had little knowledge of Washington politics, but he did know that his former Senator had been quite upset when his employer at the time, New England Electronics, failed to win the SEANET contract. He suspected that Senator Flannery would like to know what was going on at DSD, so early on Tuesday morning, precisely eight o'clock Washington time, he placed another call to the hot line number that he had obtained from Flannery's internet web site.

At the mention of the SEANET program, the call was referred to Flannery's staffer Blake Blaisedale. Although the caller did not identify himself, anonymity being guaranteed by the Senator's web site, the hot line operator recorded the call and noted the originating caller's telephone number. At Blaisedale's urging, the anonymous caller agreed to stay in touch and report any additional information that he might obtain.

Blaisedale thought the timing of the call extremely fortuitous since he had previously arranged for a status briefing from Admiral Sullivan's staff later in the week. The word of an anonymous caller was suspect and could not be taken seriously without corroboration, but after consulting the Senator, it was decided to accept the Navy briefing without them knowing that Blaisedale was aware of the replanning. If major replanning was in progress, and if the Navy did not reveal that the program was in such serious trouble as to require replanning, Flannery would have damning evidence of Rusty Sullivan's duplicity. In that case the despised Admiral would be called, subpoenaed if necessary, to appear before the Senate Armed Services Committee forthwith.

Riley's call to Washington was completed by five thirty, California time, so he proceeded with his morning routine, arriving at his DSD workstation at seven o'clock. It was a half hour earlier than his usual arrival, but he wanted to be as close to the action as possible when his next attack on SEANET occurred. He drank his usual cup of coffee in order to follow his normal routine without arousing suspicion and busied himself at his computer while keeping an eye on the clock. He knew that Jack   Decker usually arrived at DSD between seven thirty and seven forty five. Shortly thereafter there would be an unmistakable signal that Decker had arrived at his desk and turned on his computer.

Erik Hansen arrived at DSD shortly after Riley. He wanted to complete a few leftover items of paperwork before Jack Decker arrived, having left early on Friday afternoon before everything was done. His morning routine as Decker's assistant had been established over the last week, like Riley's, starting with a fresh cup of coffee. He unlocked his desk and the file cabinet used for temporary storage pending more orderly filing when time permitted, turned on his computer, and as it booted up and loaded his most frequently used software automatically, he proceeded to Jack Decker's office to make it ready for Decker's arrival. Unbeknownst to Riley, Decker did not usually come to his VP office directly but instead stopped by his SEANET workstation office first, to check his email and confer with Martin or Karinsky or both to organize the day's work.

Hansen turned on the lights that Decker preferred, opened the blinds on the south-facing windows, and closed those facing east to defeat the brightness of the morning sun. Decker, he had learned, preferred subdued ambient lighting and would use his desk lamp if better visibility were required temporarily. As Hansen departed the office passing in front of Decker's desk he reached across the broad surface and pushed the power bar rocker switch to the ON position.

The center drawer in front of the bomb became a missile that would have severed a person seated at the desk, impacting the chair and credenza behind the desk with such force that it splintered them to kindling. The drawers on either side of the desk imploded crushing their contents. But the precisely fitted mortise and tenon joints that connected the top, front and sides of the desk resisted the impact of the explosion and held the massive desk together. The entire desk as a unit was accelerated forward by the blast as if propelled by a jet engine. From his position leaning over the desk, Hansen was bent forcefully forward at the waist until his chest met the hard walnut surface of the desktop, his head thrust backward due to inertia.

His lower body struck the back of a wheeled chair at the conference table in front of the desk, which in turn, pushed the table until it struck the wall, abruptly halting the forward motion of table, chair, Hansen and desk. The energy of the blast was expended in milliseconds, and after that debris from the wooden furniture flew about wildly, papers floated to the carpet, and a cloud of smoke and dust filled the room with the unmistakable odor of gunpowder.

The explosion was heard throughout the DSD campus, the closer to the glass and brass building the louder. Riley heard it more as thunder than a crisp explosion due to echoes from various structures, but the sound was unmistakable. He looked at the clock; it was seven twenty three. To others it was a complete surprise, heard loudest by early arrivals in the executive office area nearest Jack Decker's office. The nearby security guard making his final after-hours rounds, uncertain of the exact location of the blast, was directed to the scene by the stunned early-arrivers. The guard surveyed the scene from the doorway to Decker's office, and radioed the central security station for assistance, recommending both firefighters and security. As the dust cleared he spotted Hansen pinned to the wall by the heavy furniture and could not tell if he was unconscious or dead. He radioed again, urgently requesting that medical assistance be added to the mix of responders.

Help arrived within two minutes. A nurse from the DSD medical office was among them, and after the furniture pinning Hansen had been cleared away, she determined that he was alive and breathing with difficulty, but she could not determine the full extent of his injuries. She judged them to be severe and beyond the capabilities of the first aid medical case that she had brought along. Once she determined that he was breathing, was not visibly bleeding, and had a steady pulse, she instructed the others that he was not to be moved until paramedics arrived with the proper equipment, thus minimizing the possibility of further injury. When the paramedics arrived five minutes later, they attempted to communicate with Hansen to determine if he could describe his injuries, but he remained unconscious. They carefully and expertly placed head, neck, and back restraints on him where he lie in order to protect his spinal column from further injury and to and preclude the puncture of internal organs by potentially splintered bone. When he was suitably encapsulated, he was placed on a wheeled stretcher and moved to the waiting Orange County Fire Department Emergency Response Unit ambulance at the front entrance. It was now seven thirty seven; just fourteen minutes after Riley's bomb had exploded.

Decker had not yet arrived at DSD when the explosion occurred. Just as he stopped at his alternate office, Ken Martin met him there and informed him that there had been an explosion in the executive office area. An urgent voice was then heard on the public address system throughout the DSD campus, and the voice demanded that all buildings be evacuated except for emergency and security personnel. "This is not a drill," the voice said. Decker wanted to know more and was tempted to get a first hand look, but he decided that he would probably just be in the way. Better to leave the situation to the professionals who had been trained for it. So he and Martin proceeded to the nearest exit, which in their case, was the front entrance.

Just as they approached the front entrance from the lobby, a voice called out, "Stand aside please, clear the way for the EMTs."

Decker watched as the Emergency Medical Technicians wheeled their stretcher directly past him, and despite the obscuring restraints that had been placed on the body, he recognized the tall figure with the blonde wavy hair. It was Erik Hansen.

"My God," he exclaimed, "it's Erik!" To the nearest paramedic, he asked urgently, "How badly is he hurt?"

"He's alive and stable but unconscious. Beyond that, we don't know yet," was the paramedic's terse response.

"Where are you taking him?" Decker wanted to know.

Over his shoulder, the paramedic replied, "Probably Buena Park Orthopedic, but we might be redirected."

As the paramedics lifted Erik Hansen into the ambulance, Decker, running, reached the Buick parked in his reserved space near the entrance. He intended to follow the ambulance to the hospital and get a first hand report on Hansen's condition as soon as possible. Then suddenly it occurred to him that the wrong man might be in the ambulance. Assuming that the explosion was a bomb, it was meant for him.

Although Decker was a careful, defensive driver, the Buick followed the ambulance almost bumper to bumper in order to take advantage of the path it cleared through the morning rush hour traffic on Buena Park's surface streets. Decker's experience as an amateur sports car racer, while limited, was good enough for him to keep up. As was the usual case in southern California, there were a few motorists who refused to yield to the ambulance, but the driver's experience in weaving through heavy traffic, encroaching on opposite traffic lanes or a sidewalk if required, saw both vehicles safely through to the emergency entrance of Buena Park Orthopedic. As the stretcher was wheeled into the emergency entrance, Decker parked the Buick at the nearest available spot, reserved for doctors as it happened, and ran to catch up so that he would not lose sight of Hansen. He felt responsible for Hansen's injury even though there was nothing he could have done to prevent it other than breaking his normal routine and going to the VP office first before Hansen arrived. That was no comfort.

The stretcher was met by a nurse and an orderly and wheeled immediately into a treatment area obscured from Decker's view. He informed another Nurse in the area that he knew the victim and wanted to help in any way he could. The nurse suggested that he contact the next of kin in case permission for extraordinary treatment was needed, and Decker agreed to do so. That turned out to be more difficult than he had thought since he did not have any personal information concerning Hansen, not even his phone number or his wife's name. From a pay phone he dialed Larry Hamilton's office, but received only a voice mail message. Then he remembered that the buildings had been evacuated, so he dialed the Defense Systems Division general phone number that he located in the phone book. Operators having been evacuated also, the call was automatically transferred to the main guard station, provision for which was made in order that after hours calls could be answered by the guards. Decker identified himself and asked if Mrs. Hansen had been notified. The guard answering the call did not know, but he did have Hansen's home phone number from a listing of all DSD personnel that was maintained in the security station for just such emergencies. That number obtained, Decker dialed it.

"Hello," a woman's voice answered.

"Is this Mrs. Hansen?" Decker asked.

"Yes, it is. Who's calling please?" she inquired, wary of telemarketing calls even at this early hour.

"This is Jack Decker from your husband's office. Erik has been acting as my assistant."

"Oh, yes, Mister Decker. Erik has spoken of you often – and very favorably, I might add," she responded, relieved that she would not have to hang up on another uninvited salesperson.

"I'm really sorry to tell you this, but Erik has been injured in an accident and has been taken to Buena Park Orthopedic Hospital. That's where I'm calling from."

"Oh, my God!" She exclaimed. "What happened? How badly is he hurt?"

"There was an explosion. I only know that he is breathing without difficulty and there is no sign of bleeding, but he is unconscious. I think you had better come over. Do you have transportation?"

"Yes, I do. I know just where the hospital is and I can be there in about fifteen minutes. Where shall I go?"

"Just come into the emergency entrance. I'll keep an eye out for you, and by that time we should have some better information. In the meantime, he's in good hands here. This is a great hospital, and we'll see to it that he gets the best possible treatment." Decker tried to reassure her in spite of his own concerns.

As he hung up the phone the two paramedics who had transported Hansen to the hospital emerged from the treatment room, returning to their ambulance and fire station to await the next emergency. Decker asked them how Hansen was doing.

"Hey, you're the same guy who asked me that when we picked up the injury. How the hell did you get here so soon?" the taller fireman wanted to know.

"I followed you. A little closely, I admit," Decker confessed.

"Ah-ha! You're the Buick. Usually guys that follow us like that are just trying to beat the traffic. But not to worry, we won't turn you in," the paramedic smiled. "As for your friend, not much change. He's beginning to come around, glad to say, but he's still pretty groggy. Breathing and blood pressure in the normal range. Got a pretty serious bump on the head and lots of cuts and bruises, but no serious bleeding and no apparent broken bones. They'll be taking him to x-ray after they've done some other tests and given him IVs and medication. I'm not the doctor, but I'd say he's going to be OK."

"Well, thanks very much for taking such good care of him. What station are you from?" Decker wanted to know how to contact them so that DSD could express its thanks in a more formal way. He made a mental note of it and said goodbye to them.

Monica Hansen arrived almost exactly as she had estimated. Decker had never met her, but her identity was obvious as she entered the emergency room area and looked around with concern. She was an attractive young woman in her late twenties, tall, slender and tanned with well-attended black hair falling to her shoulders. She appeared to be about six months pregnant.

"Mrs. Hansen?" Decker greeted her.

"Yes, I'm Monica Hansen," she replied and extended her hand.

"Hi, I'm Jack Decker, the one who called you."

"Hi, Mister Decker. Do you know any more about Erik?" She was obviously very worried.

Decker was glad that he could give her some assurance that Erik would recover from his injuries, repeating what he had been told by the paramedics. He also told her that he had few details of the accident, only that there had been an explosion in the office where Erik was working. "One of the nurses told me that the doctor in charge of his treatment will talk to you in a few minutes. Meantime, they have a few questions about Erik at the triage station if you feel up to it – medical history and the like I expect."

Monica Hansen thanked Decker for his kind attention to Erik as he led her to the triage station. She asked if she could see her husband, but the nurse told her that would be up to the doctor. Then she began the arduous but necessary task of filling out form after form. Decker noted that she read the papers carefully before signing them and asked questions about items unclear to her. He wondered if he could be so focused in her situation, but then concluded that this was her way of dealing with the stress of the situation.

At last a young resident in greens came out of the treatment area and asked for Mrs. Hansen. She identified herself to him, and he said, "Erik is going to be fine. He has a nasty concussion, several bruises and abrasions, and I suspect he may have some cracked ribs, but nothing serious. I've ordered a comprehensive x-ray regimen to determine if there are any broken bones or internal injuries. We should know in a few hours where we stand. If you'll follow me you can see him for a few minutes before he goes up to x-ray, but don't be surprised if he's a little confused since we've given him something for the pain."

Decker introduced himself to the doctor, asking both of them if there was anything further he could do to help – there was not – so he said goodbye to both of them after giving Monica Hansen his business card and requesting that she call him when she had a more complete report. He had concluded that Monica Hansen was in control and could manage without him.

Returning to the Buick he removed the parking ticket from under the driver side windshield wiper. It admonished him for unauthorized parking, informed him that his car would be towed upon a repeat offense, and ordered that a fifty-dollar fine be mailed to the hospital within the next thirty days. Unpersuaded that punishment for his offense rose to the level of fifty dollars, he would pay it nevertheless without objection. At the moment there were several important phone calls to make, and he reached for the cell phone that he carried in the glove box. He had not had time to have a "hands-free" installation of the phone in the Buick, intending to have it done this week or next.

The first call was to Arnie Tell's office. Susan Anders had been trying to reach him frantically but without success and was relieved to hear Decker's voice. "Are you OK, Jack?" was her first question.

"I'm fine Sue. I wasn't even in the building when the explosion occurred, but Erik Hansen is pretty banged up. He'll be OK, thank God. Does anybody know exactly what happened?" he asked.

"I'm so glad that Erik is OK. We were very worried about him – and you, too. The Sheriff's department is investigating, and they've asked for help from the FBI and ATF. They suspect it was a bomb or something under your desk since it was blown all the way across the room. Right now your office and surrounding space is being treated as a crime scene. You'll have to goof off somewhere else for a few days." She was so relieved that Decker was not hurt that she reverted to form and threw out her usual humorous jab. "Want to talk to Arnie?" she asked.

"Well, thanks for the heads up, Sue. And yes, please put Arnie on," he replied.

"Are you alright, Jack?" Tell needed to hear Decker's voice to be sure that he was not injured. Decker brought him up to date on everything that had happened since he arrived at DSD that morning, and asked if there was any more information beyond what Susan Anders had given him.

Tell replied, "All hell has broken loose at DSD as you can imagine. Security is maxed out, the Sheriff's Department is taking charge of the universe, reporters all over the bloody place outside the gates and security fences trying to get information. I've told Larry Hamilton that he should hold a press conference as soon as he gets his act together – and get the Sheriff and Mayor involved if possible. We sure don't need the publicity, so we should at least paint ourselves in the best possible light as the victim.

"But more important, right now it looks like some sonovabitch planted a bomb in your office and tried to kill you. Have you got a goddam jealous husband on your ass, Decker?"

"Not to my knowledge, Arnie," Decker offered, "and I can't think of any other reason for someone to want to get me -- nothing recently, at any rate."

"Well, we're not taking any chances. Until we get to the bottom of this, and starting right now, I'm placing your safety in the hands of Mike Harrison and his people. We'll tighten up company security at DSD to keep the sonovabitch – or sons-a-bitches – out of the place if we can, but anytime you're off campus you're going to get the full treatment. It's a pain in the ass, but you'll get used to it."

"I hardly think that's necessary, Arnie," Decker protested.

"Bull shit! This is one time you'll just have to take orders. Get your ass back to DSD, and one of my cars will be there to pick you up anytime you need to leave. So just shut up and do it." Tell was adamant about his intention to provide personal security for Decker -- even over his objections.

The call to Arnie Tell completed, Decker called Admiral Sullivan's office number from his cell phone's address book, and pressed the CALL key. When Jane Wilson came on the line, she too was glad to hear from Decker. In Washington there had been some coverage of the breaking news at DSD, but few details were available. She indicated that the Admiral was in a meeting, but she would interrupt him because he had been unable to reach Decker for their daily phone call. After being on hold for just one minute, serenaded by alternate rock music that he thought inconsistent with official business, the Admiral came on the line with, "What happened, Jack?"

"We don't know yet, Admiral. All I can tell you is that there was an explosion in my office just before I came into the headquarters building. My assistant was there and was seriously injured, but he is stable and expected to recover completely. I'm at the hospital now and haven't been back to my office, so I can't comment on the extent of the damage." Decker felt obligated to alert the Admiral that the incident might be related to SEANET, and he noted that Arnie Tell had instituted enhanced security measures to guard against additional incidents. He intended to return to DSD and would provide a full report later in the day.

With that information, even though sketchy, the Admiral decided that he should err on the side of prudence and contact the Navy security office to alert them to a potential threat. As Arnie Tell had done, he would request that the SEANET staff and work areas receive upgraded security protection until the situation was clarified.

Decker drove the Buick back to DSD, this time obeying all traffic rules meticulously, and parked in the reserved space that he had vacated just two hours previously. As he entered the lobby of the glass and brass building, he glanced at the silver Mercedes parked near the door. Inside, he was greeted by Cynthia Robbins. "Hi, Mister Decker. Remember me?"

"How could I forget, Cindy? I guess you've got the first shift to look after me," Decker replied as he extended a hand to his newly assigned bodyguard. Decker was somewhat embarrassed by having a woman responsible for his physical safety, but she was most convincing in her professional approach and sensible precautions. Understanding that her own presence would be distracting, an armed security guard under her direct supervision would be discreetly observing Decker throughout the day. It would not be necessary to be at his side full time, but any time he left the headquarters building the guard would accompany him. When he left the DSD campus, he would be driven in one of the secure company cars, such as the Mercedes parked outside, by either herself or one of the other personal bodyguards. Finally, arrangements had been made with the Sheriff's Department to patrol the area near his townhouse with increasing frequency.

It appeared to Decker that everything had been done that could be done, so now it was time to get back to work on SEANET. He walked to his VP office to see what damage had been done and what information was now available concerning the explosion. He introduced himself to the Sheriff's Department detective in charge, Henry Meisner, and was told simply, "Our bomb squad is inspecting the area now, and the ATF folks are on the way to do a more thorough examination, but I can tell you it was definitely a bomb under your desk. Somebody wanted you dead, Mr. Decker."

CHAPTER NINETEEN
INQUISITION

As the new Program Director had asked him to do, Ken Martin had made the necessary arrangements for a Subcontractor Program Review with Commware the Wednesday after Labor Day. The established SEANET practice was to alternate quarterly reviews between the subcontractor's facilities and the prime contractor's facilities, but since this was an ad hoc meeting at the special request of the new Program Director, it would be held at the Tellonics Defense Systems Division headquarters in Buena Park. Martin had requested that the Commware Program Manager, Harold Laudermilk, and his directly reporting staff be in attendance. As soon as Martin returned to his office after the Tuesday morning explosion, he contacted Laudermilk. Realizing that news of the explosion would have reached him by then, he wanted to verify that the meeting was still scheduled. Fortunately, Laudermilk had not yet left Commware's Framingham office for Boston's Logan Airport to catch his five o'clock flight to LAX. Laudermilk indicated that he had been unable to contact anyone from SEANET at DSD, so lacking notification to the contrary, he and his staff were proceeding on the assumption that the meeting was still on. Martin told him what little he knew about the explosion and indicated that business had returned to normal at DSD.

Laudermilk also wanted Martin to know that the CEO of Commware, Mason Crenshaw, was in the Los Angeles area and might drop in on the review some time Wednesday. Would he please let Jack Decker and Arnold Tell know in case they wanted to spend some time with Crenshaw? This was a subtle way of suggesting that his CEO ought to be greeted by an equivalent executive of Tellonics. Martin suggested that Crenshaw "give

Jack Decker a call this afternoon or tomorrow morning to set up a time and place."

Following his return to his VP office after security declared an all clear late Tuesday, and after his brief conversation with Detective Meisner, Decker went directly to Larry Hamilton's office to discuss the situation. Hamilton and his staff had been occupied fully with urgent decisions regarding classified material security, the safety of DSD personnel and facilities, and making arrangements to meet with the press. There was little Hamilton could add to what Decker already knew. They agreed that even though it was his office that had been attacked, Decker would remain a background figure in dealing with the media and, to the extent possible, the authorities. This was to be treated as a DSD matter rather than a SEANET matter, although they agreed that the latter was actually the more probable.

From Hamilton's office Decker placed a call to Diane Foster from Facilities. "Hey, Diane, this is Jack Decker. Looks like somebody undid all your hard work at a single blow."

"Oh, Mister Decker," she said. "I'm so glad you called and that you're OK. Is there anything we can do to get you back in business right away?"

"No, I don't think so." He replied. "I can just use the working office for the time being. I'll check it out later today to be sure everything is working. I called to let you know that it's not urgent, but when you have nothing more important to do, and after the Sheriff has released the area, you can do it all over again."

Decker's next call was to Betty Emery to give her a first hand account of Erik Hansen's condition. He was sure that the company would want to pay special attention to him and Mrs. Hansen, and Emery assured him that those actions were already in progress. "In the meantime, I'll start the search for a replacement for your temporary assistant,," Emery offered.

"You may have to throw in hazardous duty pay, Betty," said Decker. "But you said temporary, and that's appropriate. I don't know if it's possible, or if he will agree, but as soon as Erik is well enough to return to work, I'd like to make his assignment as my assistant permanent. And to hell with interviewing applicants. I'll write a job description that only Hansen can fill, including having survived physical injury from a bomb if that's what it takes."

Returning to his working office and finding everything in order there, Decker spent the rest of the day meeting with various SEANET personnel to let them know that he was not injured and that their work, like his, should continue without interruption. He promised to keep everyone informed as more information became available, and he emphasized that the company was taking every possible precaution to protect them.

When Patrick Riley became aware that Decker had escaped injury, he was furious at the failure, which he blamed on circumstances entirely beyond his control. Riley decided that in addition to his sabotage and whistle-blower activities, a more direct approach to dealing with SEANET management was in order.

Decker did not mention to others that special precautions were being taken for his own personal safety, and even though he believed them unnecessary, at the end of the day he dialed Cindy Robbins's cell phone number to arrange for his ride home. She was already in the lobby waiting for him, and they proceeded directly to the townhouse. Decker noticed that Robbins was driving faster than usual, yet cautiously, and that she was glancing frequently at the rear view mirror. Near the end of the ten-minute drive to the townhouse, the rear view glances became even more frequent; otherwise the short journey was uneventful. Decker gave her the security code required to enter the underground parking area, she dropped him directly at his parking lot entrance, and she arranged for a car too meet him Wednesday morning at seven fifteen at the same spot.

Driven to work by Jimmy Galen, another of Arnold Tell's bodyguards, Decker arrived ten minutes early Wednesday morning at the same conference room in which the Navy review had been conducted the previous week. It was his intention to greet the Commware visitors as they arrived and be introduced to them on a first name basis during morning refreshments, somewhat less formality being in order than that required with the Navy. Also, wishing to avoid any corporate level discussions, he needed to arrange for Laudermilk and Martin to attend the meeting that he had set up with Mason Crenshaw at two o'clock to make it clear that it was a SEANET only meeting. He would leave CEO subject matter to Arnie Tell. Crenshaw was the only problem subcontractor CEO that Decker had not called, so this would be the occasion for soliciting Crenshaw's support for extraordinary measures to improve his company's performance. With his own VP office now a shambles, they would meet in Larry Hamilton's executive conference room. Decker also had made an appointment with detective Meisner for three o'clock to be held in the same place. Other than those scheduled absences, he planned to be in attendance at the Commware review in order to learn first hand as much as possible about the status of the subcontract and to make inquiries as he saw the need.

Decker had instructed his own staff to make no mention of the fact that DSD was pursuing independently both hardware and software solutions to the Commware satellite software run-time problem. It was necessary, Decker, believed, to keep the subcontractor's feet to the fire searching for their own solution as the best possible course of action. If others of his staff did not do so, he would inquire deeply into this issue to see what Commware intended to do to meet its commitments. He would also raise the issue with Crenshaw if it did not come up without his prompting.

As had become their practice, Ken Martin called the meeting to order and introduced Jack Decker for a few opening remarks. Decker welcomed the Commware attendees, and then he gave a short description of the events of the previous day, pointing out that the incident was still under investigation. He was confident, he informed all parties present, that the SEANET program would not be affected at all by the loss of one Vice President's office space, a space seldom used in fact, and usually for no good purpose. The audience was appropriately amused by Decker's self-deprecating remark. Then Decker told them that he had stopped by Buena Park Orthopedic before coming to work and had spoken with Erik Hansen briefly. Other than a concussion and two cracked ribs, there were no broken bones and no internal injuries, so a full recovery was expected in a week or so. Finally, he informed the audience that he, Martin, and Laudermilk would be meeting with Mason Crenshaw in the afternoon, and it was his hope that any open issues would have been resolved prior to that time. The message was clearly, if pleasantly, transmitted: you guys straighten out this mess or I'm taking it upstairs.

Not unlike DSD's presentation to the Navy the previous week, Commware's program review was well organized, obviously rehearsed – or dry run, as such rehearsals are called – and presented with professionally prepared computer graphics displayed on the conference room's rear projection screen. As is the case in most such reviews, bad news was presented in the most favorable light possible, a practice that Decker himself tried to avoid remembering Arnie Tell's admonition, "Bad news does not improve with age."

When Laudermilk had concluded his opening presentation as a precursor to that which was to follow, it was clear that Commware had not only failed to gain ground but had slipped its schedule another two weeks. The causes of the additional delay would be reported in a subsequent presentation in the afternoon by his Software Development Manager, Jerome Levy.

Decker had also cautioned his staff not to mention the proposed delay in the satellite launch schedule, reinforcing the tactic of holding Commware's feet to the fire. 'It appears to me," Decker intruded, "that the most important thing we have to talk about today is the software development schedule. You must be aware that we're planning to launch a satellite in about sixty days, and it seems appropriate that the computers therein ought to have some software running in them when we do. So if it would not be too disruptive, could we just have Mister Levy's presentation next?"

"Well, yes, of course, if you wish," Laudermilk responded. "Give us a minute to rearrange our charts. Jerry, are you prepared to go on next?" was Laudermilk's unmistakable instruction to Levy phrased as a request.

Jerome Levy was well known to the DSD software engineers and managers who had worked with him, and they held him in high regard. Of all the Commware engineers with whom they were engaged from time to time, Levy was the most forthright, and he was a very capable software engineer with the right kind of experience for the job. He was believed to be a good manager and administrator as well. Martin whispered to Decker, aside, "He's one of the good guys, but he looks a little nervous today."

Levy skipped over a few boilerplate introductory charts and got right to the heart of the matter. The scheduled delivery date for final operational software had been slipped from October thirty-first to November fifteenth. He attributed this to some previously completed software units that had been found wanting when integrated into the next higher level of functionality. Those units would have to be rewritten, and it would take about two weeks to do it. Since they were now on the critical path of the schedule, the end date would be delayed by that amount.

"Jerry, we received your latest CSSR data on Friday, and I had a look at it over the weekend – relaxing bedtime reading, you might say," Decker said, not wishing to accost Levy personally and lightening the moment to some extent. "I didn't see the schedule slip in those data. Is this a revelation that came to light over the Labor Day weekend?" Although dismissive of some of its features, Decker had come to rely on the Cost Schedule Status Reporting system as giving a reasonable portrayal of both cost and schedule performance.

"No," Levy said. "As you know, the CSSR data lags real time events by a week or so. We became aware of this during testing just before the weekend, and I wanted you to know about it."

"I appreciate that. We do need to communicate our problems to one another directly rather than just relying on reports," Decker conceded. "But I also noted, again from your most recent CSSR data, that your labor costs in the software integration area decreased significantly during the month of August. It seems to me that since you were becoming actively engaged in software integration, the labor cost for that month should have increased. I gather that there were fewer people assigned to the job rather than more. Is that the case?"

"That's a possibility," Levy replied, the best answer he could give knowing full well that it was the case but not wishing to be untruthful. "I'll look into that and get back to you, Jack," he offered. Having heard those words before, possibly having spoken them himself in the past, Decker understood the unstated answer to be, "yes, but I'm not ready to admit it until I talk to the boss."

Then turning to Laudermilk, Decker continued, "Harold, would it be possible for you to check with the folks back in

Framingham to get us an answer before we meet with Mister Crenshaw this afternoon?"

"I'll see what I can find out," Laudermilk responded. He knew what the answer would be, and was concerned that he would not be able to reach Crenshaw to alert him to the situation before the two o'clock meeting. "If you'll excuse me, I'll just go make a couple of phone calls."

With his frequent-visitor subcontractor badge permitting him freedom of movement without escort, he left the room and proceeded to the subcontractor visitor's office that DSD maintained for that purpose. When he returned fifteen minutes later, he indicated to Decker that he would have an answer in, "a couple of hours." He had also left an urgent message with Crenshaw's assistant in the hope that she would be in touch with him before two o'clock, although she did not know his exact whereabouts and would have to rely on his calling in for messages or with instructions.

The meeting proceeded through another working lunch and into the afternoon without any additional surprises. All things considered, as software development goes, Commware was not doing a terribly bad job. Decker had seen worse – much worse. Commware's problems were not so great that they could not be corrected or worked around if the proper resources were applied, and that was the message he intended to deliver to Commware's CEO.

At ten minutes before two o'clock, Decker indicated that it was time for Martin and Laudermilk to join him for their meeting with Mason Crenshaw. He knew that Crenshaw had not yet arrived at DSD since he had arranged to be alerted in the event he arrived early. The three proceeded to the lobby to greet Crenshaw as he entered the building and pass him through security with minimal delay. Crenshaw's arrival was right on time at two o'clock, and he was greeted by Laudermilk and Martin, the latter having met him previously. Then Laudermilk introduced

Decker to Crenshaw, and after a brief and cordial exchange, they proceeded to Larry Hamilton's conference room for their meeting.

The usual afternoon refreshments had been laid out by Hamilton's assistant, and after helping themselves to iced sodas, Decker to black coffee, they took seats close together near one end of the long walnut conference table designed to accommodate ten, and they continued their informal conversation. Crenshaw wanted to know all about the explosion of the previous day, noting that the event was still prominent in television newscasts. Decker filled in what little he knew at the time and indicated that he would be meeting with the lead detective later in the day, possibly learning more. He agreed to keep Commware informed with whatever information the authorities would permit. Crenshaw expressed concern about his own facilities becoming a target and indicated that he had put in place additional security measures at the Framingham, Massachusetts, offices.

Decker having decided to let the visitor make the first move, Crenshaw at last got to the point of the meeting with the question, "How is the review coming along?"

"You will be pleased to know," Decker replied, "that your people are doing a first rate job in their presentations. I for one have learned a great deal from them today. There are some serious issues that must be addressed of course, and I'm sure that you are aware of them."

"Yes," Crenshaw agreed. "Nothing is ever quite as easy as we think it will be, is it? We have certainly had some unexpected challenges on this program. Right now it appears that the main problem is the software run time issue, but we're making steady progress."

Decker said, "Yes, I agree that's the main problem, but in the review today there did not seem to be as much progress as we had hoped. In fact, we were told that there would be an additional two-week slip in the schedule."

"Is that right, Harold?" Crenshaw addressed his question to Laudermilk.

Laudermilk, an experienced program manager, was able to disguise his discomfort. He assumed that Crenshaw's question was not a ploy and that he truly did not know about the schedule slip, so he would also not know that Decker had discovered the lack of personnel being applied to the software integration activity. Unless Crenshaw was acting, Laudermilk's attempt to reach Crenshaw with a heads-up had not been successful. Laudermilk knew that the best Commware talent, aside from Levy and a few visible others, was being applied to other programs because SEANET was not a high priority program with Crenshaw even though a major contributor to revenue and cash flow. As a cost reimbursement contract, there was little risk to the bottom line in falling behind schedule. Laudermilk saw that there was not much he could do now to protect his CEO, so Crenshaw would just have to face the music. Laudermilk was certain that he would catch hell as soon as he and Crenshaw were out of earshot. His answer, then, was most carefully phrased. "Yes, sir, that's right. Since we haven't talked in a few days, you may not know that on Friday of last week we experienced unexpected integration problems with two software units. Our best estimate is that it will take about two weeks to rework them."

"I see," Crenshaw said. "Who's working the problem?"

"Well, Jerome is on top of it, and we have the original developers back on the job as of today to redo the software." Laudermilk hoped that his answer would be sufficient to defuse the embarrassment. It was not.

"That may not be sufficient, Harold," Crenshaw admonished. "Give me a call at my hotel tonight – I'm staying at the LAX Hyatt for an early flight tomorrow – and we'll see what else might be done to help."

"I appreciate that, Mason," Decker interjected. "I can't emphasize enough how important it is to get the satellites in the air on time, and right now your software delivery is on the critical path."

"Don't worry, Jack," Crenshaw offered. "This will get my personal attention."

"There's another matter you might need to look into also, Mason." Decker intended to deliver the message clearly that in his view Commware was not doing nearly enough. "It came to light today that your software integration activity is understaffed. I can't see how you will get that part of the job done without applying sufficient resources."

In response, Crenshaw asked another question of Laudermilk. "Tell me, Harold, do you need additional people or different people to work on software integration?" Decker noted that each time he confronted him with an issue, Crenshaw had passed the buck to Laudermilk.

"Both, I think," Laudermilk responded.

"Well, let's talk about that too and see if we can improve the situation," Crenshaw instructed Laudermilk.

They went on to discuss several minor issues – again the buck being passed to Laudermilk – and there was a brief chat about prospects for future Navy business. When the meeting was over Decker was not certain that much had been accomplished. Crenshaw was certainly a charismatic figure and he appeared to be technically competent. He was being cooperative on the surface, as Decker expected he would be, but only time would tell if he was serious about assigning more and better people to SEANET software development. It was now imperative, Decker concluded, that his own DSD software people get more directly involved in Commware's activities and that the two potential DSD fixes – a faster processor and a software translator – be pursued vigorously.

The four attendees retired to the lobby, and Crenshaw departed with handshakes and warm words for all. As his final comment, Crenshaw said directly to Decker, "I'll give you a call on Friday with an update." Martin and Laudermilk then returned to the program review, and Decker to Hamilton's conference room for the meeting with detective Meisner.

CHAPTER TWENTY
SECOND TRY

Orange County is composed mostly of back-to-back, unincorporated communities stretching from the Los Angeles County line to the San Diego County line, although there is still considerable undeveloped land inland. A few of those communities have dedicated police departments, but most rely on the county to provide police services. Among other communities, the Orange County Sheriff's Department has jurisdiction over local police matters in Buena Park.

Henry Meisner became a Deputy Sheriff after having served five years as a patrolman with the Los Angeles Police Department. He had differences with LAPD procedures, was considered a bit of a troublemaker, saw little promise for future promotion, and thus decided to make the move to Orange County. Initially he served in a patrolman's capacity as a Deputy Sheriff, and in that capacity he received considerable recognition, attaining the rank of Sergeant. He had become a detective in the Sheriff's Department six years ago. When the Sheriff was informed of the explosion at DSD, a potential high profile event, without hesitation he assigned Meisner as lead investigator.

Meisner and two others had been escorted by a security guard to Larry Hamilton's conference room for the meeting with Jack Decker. Meisner introduced the older of the two as FBI Special Agent Leroy Washington and the younger as ATF Agent Pamela Robertson. They were there to make inquiries about the circumstances surrounding the explosion, in particular to find out if Decker had any ideas about who might be responsible.

After offering them refreshments, the residue of the Crenshaw meeting but still fresh and neatly rearranged by Hamilton's assistant, they assembled at the conference table in much the same manner as had the prior group of four. Decker was anxious to know what they had learned about the explosion, and he asked them for an update.

Meisner responded. "There's not much we can tell you, Mister Decker. Since we are still investigating, it's necessary to withhold certain information from the public in order to protect the investigation. I can tell you this, however. It was definitely a bomb planted under your desk. We can't release any details of its construction, but I can tell you that whoever planned this was damned clever. If you had been sitting at your desk when it went off, we would not be talking today – or any other day, for that matter. Fortunately, Mister Hansen was on the opposite side of that very massive desk, and he escaped more serious injury because the desk held together. That's all I can say for now."

"Fair enough," Decker said. "How can I help you?"

Washington posed the question. "Do you have any enemies, Mister Decker, or anyone you can think of who would want to do you harm?"

Decker replied, "Well, as you might imagine I've been thinking about that. I've probably disappointed a few people, and made a few angry, but for the life of me, I just can't think of anyone I would call an enemy. It must be that whoever did this was not after me, personally, but after the position I hold in the company. It could be a disgruntled employee, I suppose, but I've only been here a few weeks – not long enough to really make someone angry enough to want to kill me."

Meisner then said, "I understand you worked here before."

"Yes," Decker informed them, "I worked for the company from 1974 to 1990, a little over sixteen years."

Robertson then asked, "How well do you know Arnold Tell?"

"I know him quite well. We were college classmates, and we worked together on and off during those sixteen years I mentioned. You could say we are very good personal friends.

But why do you ask?" Decker was curious why his friendship with Arnie Tell might be pertinent.

"Just for background information," Robertson replied. She considered it a possibility that the bomb was somehow related to Tell, who had plenty of enemies. From that evasive answer Decker concluded correctly why she had asked about Arnie.

Meisner then asked, "Do you think you need any personal protection from us until we find the perpetrator? We can certainly do some things to improve your personal security."

"I don't think so," Decker replied. "My good friend Arnie Tell has taken care of that I believe." He went on to explain the arrangements that had been made.

"Just the same, Mister Decker," Meisner persisted, "If you don't mind we'll just keep and eye on you and your home. You won't even notice, and neither will anyone else."

"That's fine, Detective Meisner. I have no objection, and I appreciate your concern, but I really don't think it's necessary," Decker added.

"Let us be the judge of that, Mister Decker. I certainly wouldn't attempt to tell you how to build a satellite," Meisner responded.

"To tell the truth, Detective Meisner," Decker said, "I could use a little help with that just now. You don't happen to know anything about computer software, do you?" The three visitors had a good laugh at that suggestion, and after a few more routine questions, they said goodbye to Decker, letting him know that they would stay in touch.

Decker returned to the conference room where the Commware program review was just coming to a conclusion. The presentations were completed, and action items were being assigned by Martin who had, as usual, taken meticulous notes throughout the meeting, relying on Karinski's notes for the hour he was absent for the meeting with Crenshaw. Several of the participants, Commware and DSD alike, were planning to have dinner together, and they asked Decker if he would like to join them. He declined graciously, asking for a rain check, and informed them that he had a prior engagement. His engagement, of course, was the guarded ride home and an evening of dealing with the paperwork and email that had accumulated during Erik Hansen's three-day absence.

It was Mike Harrison who met Decker for Wednesday evening's short journey. Cindy Robbins and he would alternate days as Decker's bodyguard-drivers. When they met in the lobby they shook hands vigorously, recalling their pleasant evening at dinner together with Arnie Tell just over a month ago. Like Robbins, Harrison said little, was very focused on his driving, and consulted the rear-view mirror with regularity. Decker noticed several Sheriff's Department patrol cars parked along the route to the townhouse and wondered if they were part of Detective Meisner's security arrangements.

Thursday was the first routine, non-crisis day of Decker's new tenure at DSD. The only thing out of the ordinary occurred during the daily phone call with Admiral Sullivan. Sullivan had made an appointment to meet the following Thursday with Lieutenant General Joseph Scott, Commander of the Pacific Missile Test Range, who planned to be in Washington on other business. He asked Decker to join him for the meeting, and asked him to prepare a ten-minute briefing on why it was necessary to change the satellite launch schedule.

Decker gave the Admiral his impressions of the Commware program review and expressed his doubts about Crenshaw's full commitment. He was willing to give them one more chance before taking strong contractual action to get their attention. Sullivan asked if he should intervene, but Decker requested that he hold off for a while. "Maybe just a shot across the bow?" Sullivan asked.

"Let's hold off for a full volley when the time comes," Decker recommended.

Later in the morning Decker spoke with Erik Hansen at the hospital by telephone. He explained that his "chaperones" refused to let him make a personal visit there despite his protests, so phone calls would have to do. Hansen understood completely and said that he had had plenty of visitors. He was especially pleased that both Larry Hamilton and Arnold Tell had visited him together yesterday. They had chatted for half an hour before the nurse chased them out. He noted with a chuckle that Arnie Tell had made some moves on the pretty young nurse, but to no avail. Hansen said he was pretty much back to normal except for the pain in his rib cage. He expected to go home tomorrow, Friday, and wanted to return to work as soon as the doctors would permit it.

"Thank God! I'm literally drowning in paperwork. I can either do the paperwork or run the program. This job needs both of us, Erik, but don't rush it. Take care of number one first, then SEANET after that." Decker was sincere in his concern for Hansen, and Hansen knew it --all the more reason to get back to work as soon as possible.

The Thursday ride home, this time with Cindy Robbins, was just as uneventful as the last two had been. She said little, concentrating on her driving, and seemed to visit the rear view mirror less frequently than she and Harrison had done previously. The Sheriff's Department patrol cars were parked in different locations along the route but were no less in number than he had observed yesterday, six in all. Robbins slowed the Mercedes and made the right hand turn into the downward-sloping entrance to the underground parking area under the townhouse complex. She stopped before the security gate and lowered her window in order to enter the security code using the keypad on the gate control panel that extended from the ground to a convenient height for driver access. As the security gate rose and reached its halfway open level, Robbins shouted suddenly in a single breath, "GET DOWN! On the floor, Mister Decker!"

Robbins saw the silver Ford Explorer inside the parking garage on the other side of the gate. It was standing in the exit lane on her front-left about ten yards behind the gate, and at that moment it began moving toward them. She was certain that it was the SUV that both she and Harrison had been watching in the rear view mirror the last two days, but she had not seen it following them today. For confirmation she glanced at the license plate's concluding symmetrical letters, HAH. Since both front and rear license plates are required in California, she had been able to make out those three letters in the rear view mirror as it followed them on Wednesday. The danger clear and present, she reached down instinctively with her right hand for the nine millimeter "Baby Glock" automatic pistol in its holster strapped to the inside of her left ankle. As Decker had wondered when first learning that Cindy Robbins was armed, the place of weapon concealment despite the form fitting uniform was now revealed.

Robbins knew she would not fire unless she saw a weapon or was fired upon, but she would not hesitate to do so if necessary to protect her charge. As the gate neared its apogee, the SUV accelerated through the open gateway then slowed as it came up beside the Mercedes.

The weapon in the driver's hand was now clearly visible as it spewed round after round toward the rear passenger window of the Mercedes. Then, as the pistol trained toward Robbins, she fired three shots at the driver of the SUV. At least one found its slowly moving target, causing the driver to slump backward and press the accelerator pedal unintentionally. The Explorer sped forward out of the driveway until it struck a utility pole on the opposite side of the four-lane street and came to an abrupt halt, deploying its air bags. As fortune would have it, there were no other vehicles in the path of the runaway SUV at that moment.

In less than a minute a Sheriff's Department patrol car pulled up behind the SUV with tires squealing, blocking it. The patrol car had been parked a block away and rushed to the scene immediately after shots were heard. The deputies, weapons drawn, approached the SUV cautiously, shouting at its lone occupant to exit the vehicle with hands in clear view. The driver side door opened, and the driver began to step out slowly then fell out of the vehicle to the ground, now unconscious and bleeding profusely from a head wound, but still alive. The weapon was still in his hand.

Soon the street was filled with emergency vehicles. Paramedics treated the SUV driver on the scene, placed him on a stretcher, and loaded him into their ambulance for transport to the nearest trauma center. An armed deputy rode in the ambulance, a patrol car led the way, and another followed. The driver would be held under strict Sheriff's Department custody pending the formal filing of charges if he survived.

Henry Meisner, Leroy Washington, and Pamela Robertson were having dinner to conclude a long day together when Meisner's cell phone rang. "We've had a patrol report from the Decker residence. Shots were fired," the dispatcher said.

Dispatch had been instructed to contact Meisner immediately with any information concerning the DSD explosion or Jack Decker. The three sped to the scene in Meisner's unmarked patrol car with siren screaming and lights flashing. They arrived just five minutes after the shots were fired and quickly assessed the situation, relieved that Decker and his driver had not been injured. Among their first actions was the confiscation of Cindy Robbins's legally licensed Glock 17 pistol, which would be critical evidence in the inquiries that were certain to follow. They asked Decker and Robbins to wait for them inside Decker's townhouse while they gathered information outside. They would interview them there as soon as they completed their external investigation.

While they waited for Meisner, Washington and Robertson, Decker asked Robbins, "Are you OK, Cindy? That must have been very difficult for you."

"I think I'm OK," she answered. "But I sure could use a drink!"

Without asking her preference, Decker poured them both a Glenliven single malt over ice. By the time the investigators rang the doorbell, they had both had two. The visitors declined the whiskey, so Decker put on a fresh pot of coffee for them.

"Well, we know who he is," Meisner began the conversation with Decker and Robbins. "His name is Patrick Riley, and you may be surprised to know that he had with him his employee security badge from your company. Do either of you know him?"

Robbins was certain that she did not, and Decker said that he did not think so, although he might have encountered him informally in moving around the DSD campus.

"So please tell us exactly what happened," agent Washington instructed, adding, "If you don't mind, I'll just turn on this little tape machine to record our conversation." He placed the pocketsize recorder on the coffee table that they surrounded and pushed the PLAY and RECORD buttons simultaneously.

"First," Meisner added, "Miss Robbins, you fired a weapon and injured another person, so there will be an inquiry. I need to warn you that you need not make any statement without benefit of legal counsel. Do you wish to be represented by a lawyer before we proceed?"

"What do you think, Mister Decker?" Robbins inquired of Decker directly, trusting in his judgment rather than her own.

Decker offered his opinion. "I think it's alright for you just to tell them exactly what happened, Cindy. I'm not a lawyer, but I don't think you are in any legal jeopardy."

Then Robbins described the incident as she recalled it, pointing out that both she and Mike Harrison had observed the silver Explorer following them on Tuesday and Wednesday and that she had observed the partial license plate number, HAH, which looks the same even when viewed in a mirror. When she saw that vehicle with its telltale license plate waiting for them, her suspicions had caused her to warn Decker and to draw her weapon in case it was needed, and as it turned out, it was.

Decker's account was in perfect synchronization with Robbins's from his point of view. Upon her warning, he had ducked to the floor as instructed, trusting her judgment then as she was his now. The next thing he heard was five or six muffled shots fired in rapid succession and a loud crackling sound each time the bulletproof glass was struck. Then he had heard three additional shots with a loud report at close range. He noticed when he sat up, he said, that while the glass had shattered, it remained intact and was not penetrated. "I guess I owe one to Arnie Tell for insisting that I take security precautions, and to Cindy here, for reacting so well in a dangerous situation."

The interview was soon completed, with the understanding that Decker and Robbins might be questioned further at a later time. The Mercedes and Robbins' baby Glock automatic had been impounded as evidence, Meisner told them, so he offered to have a patrol car take Robbins home. When Robbins headed out the front door of the townhouse, she turned and gave Decker a warm hug and a kiss on the cheek. "It was real scary," she said, "but we came through it OK, didn't we?"

"Yes, we did, thanks to you. I'll always be gratefully, Cindy." Decker participated enthusiastically in the embrace of the brave and beautiful young woman, intended as a warm measure of comradeship and not interpreted as anything more by either of them.

After she left, Decker placed calls to Susan Anders, Arnold Tell, Larry Hamilton, and Ken Martin, in that order, and he

242

reached them all at their homes. Each conversation began with, "Guess what happened to me on the way to the townhouse!"

Harold Laudermilk called Mason Crenshaw at the LAX Hilton as soon as he got back to his own hotel from the Commware-Tellonics dinner. Crenshaw was not in his room, so Laudermilk left a message, and his call was returned just after ten o'clock. There was no exchange of personal greetings, Crenshaw lighting into Laudermilk without hesitation. "What the hell do you mean letting that sonovabitch Decker sandbag me like that? Why wasn't I told that he knew about the staffing problem, and how come you slipped the schedule again without my knowing about it?"

Laudermilk explained that he had tried to reach him earlier in the day without success, but Crenshaw's wrath would not be assuaged. When the conversation concluded, Laudermilk had no doubt that he was in serious trouble with the CEO. He hoped that he and Crenshaw would not be on the same flight to Boston tomorrow morning because he wanted no repetition of the tongue-lashing just delivered. It happened, however, that they were on the same flight. Hoping to avoid Crenshaw, notoriously a last-minute boarder, Laudermilk, with frequent flier privileges, was among the first to board and took his seat midway into the coach section of the Boeing 757. He was not seen by Crenshaw who entered later and sat forward in the first class section. After the Boston landing, Laudermilk waited until Crenshaw had deplaned and was well on his way before he left the plane himself.

Fearing the worst, Laudermilk did not drive directly to his home in Sudbury. He went instead to the Commware office building in Framingham. The guard on duty, who recognized him and checked his security badge, allowed him to enter after he signed the after-hours log. During the five-hour flight from Los

Angeles he had decided what steps had to be taken to protect himself, and he set about the task.

First he copied his critical computer files onto removable Zip disks. His invariant practice was to load all operating system files and application programs on his primary hard disk drive, the logical "C" drive, and all data on a separate hard drive identified as the logical "D" drive. The important data were copied from the "D" drive and they included all files related to SEANET, which he kept in separate directories by category, his recent and archived email files from the last five years, and files that related to personnel matters since joining the company. When he finished, an hour and a half had passed, and he had filled twelve Zip disks holding almost one hundred megabytes each.

That completed, he proceeded to clear the "D" drive of any trace of its data. He knew that the "delete" function did not actually remove data from a hard drive but only changed the identity of the files so they would not be recognized by applications and could be overwritten, but they could still be recovered with special software. To truly make his data on the hard drive non-recoverable, he ran a utility program that wrote "ones" over the entire extent of the disk followed by a symmetrical pattern of "ones" and "zeros" leaving no trace of prior content. If things worked out, he could restore his important files by copying the Zip disks back to the "D" drive. If not, only he would have access to the critical data it once contained.

Next he searched his desk, bookcases, and file cabinets for hard copies of data that either he wished to retain or that he did not want to leave behind for others to see. There was some such, but not much, since he relied almost exclusively on his computer and seldom printed other than general reference documents. Finally, he checked his classified data file cabinet to be certain that all documents listed on the inventory were present. Then he went home, his briefcase heavier than it had been when he arrived.

As Laudermilk approached his office Friday morning, he saw taped to the door a terse note from his boss, David Crawford, Vice President and Director of Software Program Management. "Harold, Please stop in right away when you arrive. Dave." After dropping off his briefcase, now lighter without the material deposited there last night, and after filling his cup with black coffee, he went to Crawford's office and was invited to "go right in" by his assistant. Not only was Crawford present, but also Frank Diamond, Vice President and Director of Human Relations. There was also a security guard standing in the far corner of the room. Laudermilk shook hands with Diamond, whom he had met on several prior occasions but not seen lately, and he said hello to Crawford, then he nodded to the security guard with a smile. His intuition had been correct; this was to be his last day as an employee of Commware. He wondered only how his departure would be characterized and explained.

When they were seated, Crawford was the first to speak, his words obviously planned carefully in advance. "Harold, we have come to the conclusion that it would be best for all concerned if you left the company. We would certainly like your departure to be as amicable as possible, and so we would prefer that you resign with no hard feelings by either party. Frank, here, can explain the generous severance package he has arranged that will be more than adequate to tide you over until you find employment elsewhere. This must be a surprise to you, but we'd like to have your resignation effective today."

After a moment's hesitation as he considered the circumstances, Laudermilk responded. "I'm curious why you think it would be in my interest to resign, and I certainly don't agree that it would be in yours. But my real question is: what are the alternatives if I don't choose to resign."

Diamond answered that question. "The alternative, and there is only one alternative, would be less pleasant for both

246

parties, Harold. Termination for cause. In that event, few if any benefits would accrue to you, and you would not be able to obtain favorable recommendations from Commware."

Laudermilk's response was, "May I ask what the 'cause' of my termination is?"

As they ping-ponged the answers between them, it was Crawford's turn, and he replied, "We are not required to provide specific causes to you, but in fairness let's just say that we are not satisfied with your recent performance. It is sufficient to indicate that you will be terminated at the discretion of management."

"I see," Laudermilk said. "Well, I choose not to resign, so you'll just have to fire me, Dave. What's the next step?"

Diamond took it from there. "I have here a termination letter addressed to you. There are two copies. One is for you, and I'd like for you to sign our copy to acknowledge that you have received it. Then we will go over to your office so that you can collect your personal effects and turn over to us any keys, identification, or other company property in your possession. And since there are classified data in your care, a Security Department representative will conduct a debriefing, and your classified documents will be inventoried and returned to central files. After that, we will say our goodbyes at the front entrance."

Laudermilk said, "I'd prefer not to sign anything, if you don't mind, Frank. The three of you are witnesses to my dismissal, and that should be sufficient without my acknowledgment. On the other hand I would like receipts indicating that I have returned all company property and that my classified documents are in order; I think that's routine security practice. So shall we just move along and do what has to be done?"

+++++++++++++

The phone rang repeatedly throughout the morning as Decker's friends, acquaintances, and business associates inquired as to his well being, and to get a first hand report of the previous night's excitement. Among the calls were two from reporters, and Decker gave them the agreed upon "no comment," referring them to the Sheriff's Department. He soon tired of the calls, and was about to leave for an early lunch when the phone rang again. Decker said to himself that this would be the last one he answered this morning. It was Harold Laudermilk. The second attempt on Jack Decker's life had not yet been associated by the press with the DSD bombing incident and had not been widely reported on the east coast. Laudermilk not being aware of it, this call did not concern the shooting but rather Laudermilk's personal situation.

"Well hi, Harold. How Goes?" Decker inquired.

"Not very well, I'm afraid," Laudermilk answered. "I got fired this morning."

Decker was shocked. "That's bad news, Harold. I thought we were just beginning a good relationship. What the hell happened?"

"Guess I must have pissed off Mason Crenshaw, and he didn't waste any time getting rid of me. He was pretty upset when you brought his attention to the manpower and schedule issues, and I had not been able to give him a warning that you would bring up those issues. But I think it goes a lot deeper than that. I believe he is concerned that I have access to information that would be embarrassing to him and Commware." Laudermilk did, indeed, have such information, but he had kept it to himself.

Decker replied, sympathetically, "I can understand that he might be a little miffed. Nobody likes to be caught short. But that's no reason to fire a good employee. And as for embarrassing information, every company has a skeleton or two in the closet."

"Not skeletons that involve business ethics, violated contracts, lies told to customers, and perhaps even laws broken," Laudermilk offered.

"What do you plan to do now, Harold?" Decker wanted to know.

"Two things," he replied. "I'm seeing an attorney Monday to institute an unlawful termination suit. And I want to give you some information you need, but it has to be in person. When and where can we get together?" Laudermilk had decided to make Crenshaw regret his precipitous action, and to that end he would share with Decker all he knew about Crenshaw's business practices. Decker would be the conduit through which adverse information would do Crenshaw the greatest possible harm.

Decker said that he expected to be on the East Coast the following week and would get in touch with him to arrange a meeting.

"The sooner the better," Laudermilk said. "I've sat on this stuff entirely too long out of loyalty, but now it's burning a hole in my pocket – and my conscience." He gave Decker his home and cellular telephone numbers and made it clear that he was anxious to get together as soon as possible.

Decker's Friday had begun when Mike Harrison met him in the townhouse parking garage at seven o'clock in a security-enhanced Lincoln town car. He told Decker that he had spoken at length with Cindy Robbins Thursday night, and had arranged for her to take a few days off to come to grips with having had to fire her weapon and injure another person even though in self defense. If she needed counseling, the company would provide it. In the meantime, it was not clear that Patrick Riley had acted alone, so until that was established one way or the other, the special security arrangements for Decker would continue. If Robbins were not ready to return to work on Monday, then he, Harrison, or another member of Arnie's army, as Harrison called it, would meet him by prearrangement on Monday or over the weekend if he needed transportation.

Decker met first with Martin and Karinski, as was his usual practice. He gave them a full accounting of the events of the night before, and then he requested that they organize a meeting of the SEANET staff at nine o'clock. When Decker mentioned the name, Patrick Riley, both Martin and Karinski said that they knew of him but had never worked with him personally. Decker hoped that other members of the staff might be able to shine some light on Riley at the nine o'clock meeting.

Admiral Sullivan reached Decker for their daily phone call and was alarmed by Decker's recounting of last night's events. They concluded that the press had either not heard about the shooting or had not associated the two attempts on Decker's life. They would soon enough, they expected, so the media would eventually make the connection with SEANET. They agreed on a "no comment" strategy. They would refer all questions to the Orange County Sheriff's Department with the explanation that the investigation was still in progress and they were not at liberty to discuss it.

As soon as the Sullivan call was completed, Decker dialed Detective Meisner's number. He wanted to know what if anything had been learned about Patrick Riley.

Meisner said, "Listen, Jack – is it OK if I call you Jack? This 'Mister' and 'Detective' stuff is getting a little old. I'm just Hank, by the way." Decker agreed, and Meisner continued. "I can't release all that we've learned, but since you were the intended victim, you deserve to know at least some things. Can I count on you to keep this to yourself?"

Again Decker agreed, and Meisner continued once more. "Your lady bodyguard turned out to be a pretty good shot with that little pistol of hers. Riley took one in the left shoulder, and another fractured his skull above the right eyebrow, just missing brain tissue. The guy lost a lot of blood and is still in intensive care, but they think he will survive without permanent damage. We got a warrant to search his house last night. He had a substantial cache of weapons, and we found bomb-making material that was identical to that used in the bomb that exploded in your office. ATF is working to nail that down. And would you believe he shot at you with an old German Luger? We could never have traced it, except it was still in his hand when we captured him. Well, anyway, the FBI checked out his background, and their profiler is convinced that he was a loner with no terrorist associations. The usual thing – a lonely guy looking for recognition. So at this moment it appears that he's the one who planted the bomb and that he acted alone. What we don't know is why, so we'll need to talk with you and others at your company to see what we can learn about that."

Decker offered to help in any way that he could and mentioned that he would be meeting with his staff shortly. If any of them had information about Riley, he would put them in touch with Meisner.

Decker opened the staff meeting with a brief description of the second attempt on his life without disclosing the details he had received from Meisner. He pointed out that over his objections Arnold Tell had had the foresight to insist that he be guarded when traveling. He heaped praise on Cindy Robbins, ending with a warning not to try to take advantage of pretty blondes in chauffeur's uniform. Finally, he identified his attacker as Patrick Riley, asking the group if they knew him.

Lee Chang spoke up at once. "I just can't believe that it was Riley. He's a really quiet guy who never makes trouble, and he does first-rate work. He's a design draftsman in my group responsible for the CADAM drawings for SEANET microwave components." Chang referred to the Computer Aided Design and Manufacturing drawing system, CADAM, used throughout DSD for creating and maintaining engineering drawings and production documents required by manufacturing. The engineering design and the manufacturing processes were automatically linked by this system, making the transition from design to manufacturing much more reliable than that provided by hard-copy drawings.

"How long has he been with us?" Decker wanted to know.

"Not very long," Chang replied. "He worked for us as a job shopper for about six months, and his work was so good that we took him on permanently a couple of months ago. Before that I recall that he had worked in Massachusetts – for New England Electronics, I believe."

With this revelation, two of those present had somewhat different but simultaneous flashes of inspiration. Manufacturing's Ralph Brown asked, "Was he responsible for the microwave board drawings?"

As that question was asked, Decker remembered the anonymous mystery note he had received two weeks ago: "The answer to your problem is well documented."

Chang answered Brown's question. "Yes, he was."
Brown had suspected all along that there were microscopic dimensional differences between the engineering prototype boards and the production boards, but it was impossible to detect them by ordinary means. He had suggested to the graybeards investigating the microwave circuit board problem that they consider that possibility, but they had not yet gone down that difficult path. Brown asked, "If this guy was crazy enough to explode a bomb and then try to assassinate the boss with a gun, is it possible he could have screwed up the CADAM drawings somehow?"

Chang said he doubted it. Drawings could not be changed without the responsible engineer's approval. Nevertheless, interpreting the mystery note as pointing to a documentation problem, Decker asked, "How do we know he followed the rules?"

Chang reconsidered. "Well, I suppose we don't know with certainty. He could have changed something without consulting an engineer, but that would automatically trigger a change to the next revision letter of the drawing in the CADAM system. That can't be done without the engineer's concurrence and password entry. Unless, of course, he found a way to work around that check in the software or got the password somehow. Even so, someone would have noticed that drawing revision levels had changed. It would be pretty difficult to do."

"Difficult, maybe, but impossible, no. Let's take a very close look at that. Get the graybeards on it, and have an expert from the CADAM software company come in and take a look at Riley's workstation to see if there's been any funny business going on there." Decker saw this as a very likely possibility and wanted maximum effort applied to either verify or discount it. "I'd like your report on progress, Lee, until we get to the bottom of this."

Brown offered to assist in any way he could. He suggested to Procurement Manager Bob Masters that he have the circuit board vendor examine the data electronically transmitted to them from DSD's CADAM system. At the vendor's computers, the data could be viewed as "from-to" lists and precisely defined geometric coordinates of conductors and connection nodes on the individual layers of the boards. Better than anyone else, they could detect microscopic differences between the prototype boards and the production boards. Decker endorsed that as a promising approach, and Masters promised to get action started before noon.

Decker asked Chang to call Detective "Hank" Meisner at the Sheriff's Department and pass along all that he knew about Riley. Finally, since the staff was assembled, he took the opportunity to go over a few other current issues. The meeting was over in half an hour and he returned to his office to take care of paperwork, email, and phone calls – including the extraordinary one from Harold Laudermilk.

+++++++++++++++++++++++++++++++++++

After returning from lunch, Decker, regretting Erik Hansen's absence, settled in to clear up the remaining paperwork and to check for new email and telephone messages. One of the voice mail messages was from Mason Crenshaw, calling on Friday as he had promised to do in their Wednesday meeting. Decker determined that he would not let on that he knew about Laudermilk having been fired, then returned that call first as the most important of the lot.

When Crenshaw came on the line, not aware of the previous day's shooting incident, he was at his exuberant best. "I discussed the manpower issue with Harold Laudermilk, and I gave him hell for not letting me know about it sooner. But never mind that, the important thing is that we will be shifting quite a few experienced software integration people onto SEANET within the next two weeks. Some of them will start on Monday, actually. So I think we can bring that under control right away.

"The second issue, the schedule, is not so easy to solve. I did verify with Laudermilk and Jerome Levy that there could be an impact due to a couple of misbehaving software units, but I think Levy has overstated it. We're looking into some workarounds.

"Levy also told me that you were concerned about having software available for the satellite launch in November. I have no doubt that we will solve this timing problem eventually, but I can't guarantee when. So what we propose is that we launch the first satellite with a "SEANET Light" software package. It would be fully functional, but limited in capacity and without a few bells and whistles. Good enough to get the test program started. Then, as fixes accumulate, we can upload them to the satellite. By the time full capability is required, we should have a fully compliant software load. What do you think?"

Decker's first thought was that Crenshaw was not only deceptive but an outright liar. He must have known about, had probably even been responsible for, the manpower shortage. And as for his having spoken to Laudermilk this morning, that was just a lie. Nevertheless, he had to respond to Crenshaw's proposal. As a last-ditch, drop-dead solution, it might be acceptable, but the situation was not that desperate – at least not yet.

"That's one possible solution, Mason, and pretty clever at that," Decker replied, "but I'm concerned about diverting resources to developing SEANET Light, as you call it. Right now I think you should concentrate on integrating the mainstream program and working with the Ada experts to speed things up. We have a couple of things in the works here that might help, and we think that some of the Ada constraints can be safely no-opped." Decker used the industry jargon, no-op, to mean software routines that could be bypassed and made non operational. "Milt Karinski will be working with Jerome to coordinate all of this. So hold off on SEANET Light, but keep it on the back burner in case we come to that.

"But we can talk more about it next week if you're available. I have to be in Washington next Thursday, so I could stop by Framingham on Wednesday if you're available. I'd like to see your operation first-hand anyway." Decker thought it imperative that he talk with Crenshaw face to face, possibly confronting him with hiding something, but he wanted it to appear to be a routine courtesy call.

"Let's see," Crenshaw said as he reviewed his computerized calendar. "I've got a free hour at ten o'clock, and after that you can have the grand tour. Delighted to have you if that's satisfactory."

"More than satisfactory," Decker said. "See you Wednesday at ten. And have a great weekend, Mason."

"Same to you, Jack," Crenshaw replied.

Decker did not know exactly what might develop during his meeting with Crenshaw. It would depend largely on the information he received from Laudermilk and coordination with Arnie Tell.

As soon as he hung up with Crenshaw, he punched in Laudermilk's cell phone number. Laudermilk confirmed that he and Crenshaw had not met that morning. He agreed to meet Tuesday night for dinner after Decker arrived in the Boston area.

The next call was to Arnie Tell. After a lengthy, but innocent, exchange with Susan Anders, Tell came on the line. "You still here, Decker?"

"I suppose I should take the hint. Somebody around here doesn't like me very much. Now we know the guy's name, and it turns out that he is one of us, Patrick Riley, a design draftsman in the Microwave Engineering Group. And it's possible that he may have done something to CADAM data to screw up four circuit boards that have mysterious production problems. God knows what else he might have done. But that's not the main reason for my call." Decker went on to describe his conversations with Harold Laudermilk and Mason Crenshaw.

Tell's reaction was, "I always figured the guy to be sleazy. Now that we know for sure that he's a liar on small things, he probably lies about bigger things, too. Before either of us talks with Crenshaw again, we have to balance the SEANET interests with the corporate interests, so we'd better get together before you leave for Boston to decide how to handle him. Problem is, I'm really booked solid today and Monday. In fact you caught me between meetings or we wouldn't be talking. How about dinner Monday? You can stay overnight at the airport rather than driving back and forth to Buena Park. Sue can set it up and let you know when and where. We'll go someplace expensive, and this time you pay, Decker!"

Decker agreed, and Tell returned to the previous subject. "I spent a little time with Cindy Robbins. I was concerned that she might have some problems with the shooting, you know, like cops do after they shoot somebody. But she seems to be OK with it. One helluva broad, you gotta admit looks, brains, balls – all in one package. You still looking for a woman, Decker?"

"Thanks for the suggestion, Arnie, but Cindy and I already have a relationship – a close but platonic relationship. Besides, I'm OK in that department."

"Is that so?" Tell prodded. "We'll have to talk about that, too, Monday night."

"Not on your life, Arnie. Military secret."

CHAPTER TWENTY THREE
STRATEGIES

Senator Flannery's staffer Blake Blaisedale attended the regularly scheduled meeting with Admiral Sullivan's SEANET staff on Thursday afternoon, the day of the second attempt on Jack Decker's life, that event yet to occur due to the time difference between east and west coasts. As was their custom, Sullivan's staff presented the current status of the program and the outlook for the future, with emphasis on the coming month. The SEANET staff considered the meeting a nuisance but recognized it as a necessary part of congressional oversight of executive agencies. Staffers from both the House and Senate Armed Services Committees attended, and occasionally an actual Senator or Representative would appear, but that was a very rare event. The politicians preferred to conduct their meetings on their home turf in the Capitol building's hearing rooms. At this meeting Blaisedale expected to hear about major program replanning as confirmation of his anonymous informant's information -- the informant having been identified as Patrick Riley by surreptitiously tracing his several phone calls over the last two months despite guarantees of whistle-blower anonymity.

Admiral Sullivan had instructed his staff not to present the new schedule proposed by Tellonics since it had not yet been agreed to by the Navy, but also not to deny that the schedule was being reviewed should the congressional staff inquire about it. It was Sullivan's intent to go public with a revised program plan only after his staff had completed its analysis and all necessary bases had been touched – such as the satellite scheduling base to be touched in the meeting next week with General Scott from Vandenberg. Without realizing it, Sullivan's strategy played directly into the hands of his most ardent critic. So at the conclusion of the meeting with Sullivan's staff, Blaisedale returned to the Capitol to debrief Senator Flannery.

"They did not say one word about a new program plan," Blaisedale reported to the Senator, "even though it was obvious from their CSSR data that the program was both over budget and behind schedule."

"How did they explain that?" Flannery asked.

"They didn't, and I didn't ask them to. They just put the data up and moved on to the next subject. I think the other staffers were either too dumb to catch it or didn't give a damn, so they didn't ask either." Blaisedale held his fellow staffers in disdain, not without considerable justification based on their past performance.

"Well I'll be damned." Flannery was delighted. "I think we've got the bastards just where we want them. Let's see if we can get them to tell the same lies under oath. I want Sullivan in here next week, and let's get that new wonder boy program manager from Tellonics in here, too."

+++++++++++++++++++++

Jack Decker returned to his office on Monday morning refreshed after a pleasantly uneventful weekend. Because of the strict scrutiny of his bodyguards and their admonition not to travel by means other than their security-hardened vehicles, it was not possible to spend the weekend at the beach house, as he would have preferred, lest his relationship with Susan Anders be disclosed. Susan, however, was under no such constraints, so she drove to the townhouse late Saturday afternoon with the ingredients for a very nice dinner, and she departed Sunday afternoon. At least they were together for a day, even if not free to leave the townhouse together.

Susan had insisted on making his airline and hotel travel arrangements for the coming week through the Tellonics travel agent -- except for the secure ground transportation, which Decker had arranged himself with a phone call to Mike Harrison. Some of the limousine trips were just minutes long, covering a few blocks, but Harrison insisted: only secure limousine or commercial airline transportation until further notice. Even though Patrick Riley was in custody, it was not yet established that he had acted alone, so Harrison was taking no chances. The schedule: seven thirty Monday morning limo to DSD Buena Park office, six o'clock limo to West Hollywood for dinner with Arnie Tell followed by limo to LAX Marriott for overnight stay, seven thirty Tuesday morning limo to terminal five for American Airlines flight to Boston, limo to Logan Airport Hilton for check in, six thirty limo to Boston for dinner with Harold Laudermilk and limo return to Hilton, eight thirty Wednesday morning limo to Framingham for Commware visit, three o'clock limo to Terminal B at Logan airport for US Airways shuttle to Washington, limo to Arlington Hyatt for Wednesday night stay, seven thirty Thursday morning limo to National Center 2 in Crystal City for meeting with Admiral Sullivan and General Scott, one o'clock limo to Dulles airport for three o'clock United Airlines flight to LAX, six thirty limo to the Buena Park townhouse. It was all carefully planned. On the east coast a security contractor would provide the limousine and bodyguard service, and on the west coast, Arnie's Army would do the honors.

This was to be a week of meetings, starting Monday night with Arnie Tell and ending with the customer in Arlington on Thursday. Meetings were condemned by some as a waste of time and were the subject of much corporate humor, but Decker had found that they were often the best way to communicate with peers, subordinates, and superiors in the organization. True, some meetings were wasteful of time and intellectual energy, but any waste in a well-organized meeting was more than offset by the efficiency of simultaneous communication of information to a group whose members needed to proceed in coordination. An interesting by-product of electronic mail, Decker had also found, was that it was another means of communicating simultaneously with members of a group, so that the need for meetings had decreased somewhat since it came into widespread use. Nevertheless, there was no substitute yet discovered, even video teleconferencing, to take the place of person-to-person, eyeball-to-eyeball meetings between interested parties, be it a party of two or a dozen. Beyond a dozen, in Decker's view, meetings did not result in the free flow of information among parties present, but rather the few presented their thoughts to the many on a one-way street. Other than the planned tour of the Commware facilities, this week's meetings were to be of the former, efficient category.

Decker had packed for the trip Sunday night after Susan Anders' departure. He preferred to travel light, just a small, black, rolling case that could be hand-carried aboard the airplanes along with the matching laptop computer case that doubled as briefcase. For this three-day trip, the navy-blue blazer he wore was the only jacket required. It would be carefully hung when not actually in use to avoid wrinkling, a characteristic of the expensive fabric selected specifically for that quality. The suitcase contained shirts, ties, socks, underwear, handkerchiefs, two pairs of trousers, black dress shoes (brown loafers would be worn) and his irreplaceable "health kit." It was a small, transparent, soft plastic case with several compartments that contained everything likely to be needed when traveling. Almost all of the items were miniature versions that had been accumulated over years of domestic and international travel, many acquired to treat one malady or another encountered, and of course, the usual toilet articles, also mostly miniatures. If necessary, Decker could travel indefinitely with this light baggage by using same-day hotel laundry and cleaning service, and since it was all "carry on," there would be no waiting for baggage at destinations.

After the morning telephone conversation with Rusty Sullivan, Decker proceeded to closing out unfinished email business and meeting briefly with various members of the SEANET team to check on this point or that. He accumulated the documents he would need to take along on the trip, including the ten-minute briefing charts for General Scott prepared for him by the graphic arts staff, and was thankful that they all fit into the computer case, if barely. He called Erik Hansen, now released from the hospital and resting comfortably at home, to check on progress. To Decker's great relief, Erik expected to return to work the following Monday.

Ken Martin dropped into Decker's workstation office at three o'clock with both good news and bad. Decker opted for the good news first – he always selected that option. "You were right. There is something wrong with the microwave circuit board drawings. One of the graybeards was using some old hard copy drawings that he had in his files from when he had participated in a design review of the boards about six months ago. Another guy who had not worked on them before had ordered some drawings from the print room, so of course, he got drawings to the latest revision level. When I mentioned to them that Riley might have done something to screw up the drawings, they compared notes. Sure enough: on several drawings there were small differences between the old and new drawings in some of the stripline dimensions. Now that's not so unusual; drawings do get revised in the normal course of events. But the older and current drawings that were different had the same revision letter. So we know that something went wrong in CADAM."

He was referring to the Computer Aided Design and Manufacturing computer software used for creating engineering drawings and communicating that information directly to computer programs used by the manufacturing and purchasing departments. He continued, " Changes were made on drawings and approved without the revision letter being upgraded. And at these ultra high frequencies those stripline dimensions, as you know, are critical to efficient power transfer from one circuit element to the next. The graybeards tell me that the differences were small and not obvious without microscopic inspection, but they may be sufficient to cause the problems we've been investigating.

"Damn!" Decker exclaimed. "That bastard caused weeks if not months of delay with the stroke of a pen – or a keyboard, I should say. Well, at least we have a clue. And what's the bad news?"

"We can't figure out how he did it. The CADAM software vendor came in first thing this morning, and he's baffled so far. So the problem is this; we know he screwed up these drawings, but there may be others and we don't know how many." Martin saw this as a more pervasive problem than had been thought originally.

"Can you put together a list of all the drawings he's worked on?" Decker inquired.

"Probably. Then we'll have to check every damn one of them. There could be hundreds. It could take weeks, and even then we could miss something. Unless the software guys can come up with something specific, it's a needle in a haystack problem." Martin was not optimistic.

"Well, there's another possibility. Riley is still alive. Maybe the police can get something out of him when they are able to question him. I suggest you give Detective Meisner a call and ask him to look into it." It was the only thing that Decker could think of to do, but he held out little hope for a positive response. "I'm traveling tomorrow, but maybe there will be some answers when I return on Friday. Meantime, Ken, hang in there. Every problem has a solution. We just have to find it."

After Martin left just after four o'clock, there was another phone call. "Mister Decker, this is Audrey Nelson at the reception desk. There's a United States Marshal here to see you on official government business. He asked if you could meet with him for a few minutes in the lobby." Miss Nelson concluded, as did Decker, that the marshal's visit was related to the bombing incident.

"Mister Decker?" the man asked as Decker entered the lobby.

"Yes, I'm Jack Decker," was the reply.

The marshal reached into his inside coat pocket and extracted a folded document that he presented to Decker, and Decker instinctively reached for it. "This is a subpoena for you to appear Thursday before the Senate Armed Services Committee in Washington. If you have any questions, you can call any of the numbers listed in the accompanying fact sheet. Thanks for your time." With that the marshal, mission accomplished, turned and departed with no further comment.

Decker, surprised but calm said nothing, turned away, and returned to the workstation office where he placed two calls. The first was to Admiral Sullivan at his home, the number given him for emergencies and called for the first time on this occasion. "I thought you should know I've just been subpoenaed by your ol' buddy in the Senate. Any idea what this is about?"

"Sonovabitch!" was the response from the Admiral. "I have no idea what they are up to. I'm scheduled to meet with the committee on Thursday morning after our meeting with General Scott, but that's just routine. I go over there for my periodic attitude adjustment session every month or so, but this is the first time I've heard of a contractor program manager being called up out of the blue. I'll see what I can find out and call you tomorrow."

"I'll be at the LAX Marriott until seven thirty tomorrow morning, after that I'll be traveling. If you can't reach me, please contact Arnie Tell's office. I'll stay in touch with them for any messages, and I'll call you later in your day. You may want to discuss this with Arnie anyway, and I'm seeing him in a couple of hours, so I'll discuss it with him also." Decker wanted to be sure that he, Sullivan, and Tell were absolutely on the same page.

The second call was to Arnie Tell's office. Susan Anders answered and informed Decker that Tell had departed at noon for outside meetings and she could only leave messages for him. If he called in before their scheduled dinner, she would ask him to call Decker's cell phone. Then, in coded language they had developed to pass personal messages disguised as business conversation, she concluded with, "It's been a very long day." Translated, the message was, "Call me at home later."

++++++++++++++++++++++

Mike Harrison picked up Jack Decker in the Glass and Brass lobby at six, and they drove directly to Mirabelle's in West Hollywood, a less-well-known but excellent restaurant -- and expensive as Arnie Tell had demanded since it was Decker's turn to pay for the dinner. Mirabelle's sat almost directly across Sunset Boulevard from Spago's, the restaurant of international fame and the place to be seen if one were in the entertainment industry. Local humor had it that only those who were unable to obtain a table at Spago's ate at Mirabelle's, but Tell preferred it for its quiet ambiance and lack of celebrity. Movie stars could be seen there, too, from time to time, but no particular fuss was made over them as both they and the management preferred. Harrison escorted Decker inside and accompanied him and the maitre d' to his table, and after surveying the situation, he waited outside for Tell, who arrived about ten minutes later in a separate Tellonics limousine.

When the three of them were seated, greetings exchanged, and cocktails ordered, Decker began the conversation. "I guess you're going to be a little jealous, Arnie. Late this afternoon I was served a subpoena from the Senate Armed Services Committee."

"You're shittin' me!" Tell exclaimed.

"No, it's true. This Thursday. I talked to the Admiral and he has no clue what it's all about," Decker said.

"I'll be damned. Well, we need to talk about that in addition to all the other crap that's going on. And yes, I am jealous – or at least confused. Why did they call you instead of me -- or either one of us for that matter? They must be digging for specific information on SEANET, and that means that sonovabitch Flannery is behind it." Of the latter, at least, Tell was certain.

"Sullivan will find out whatever he can and call us in the morning. If he can't reach me, he'll call your office. It turns out he'll be testifying also, but in the morning, so I'll be seeing him before I go on at two o'clock. Screws up my reservations, of course, but if it's in the national interest, what the hell?" Decker considered this a serious matter, but by making light of it he hoped to defuse it.

"Well, you'll need to talk to the lawyers, of course. We keep a bunch of them on the payroll for just such eventualities, so we might as well have them do something useful for a change. I'll alert Ray Sycamore, and you can call him sometime tomorrow." Tell referred to Tellonics' Vice President and General Counsel, Raymond R. Sycamore, Esquire, as the lawyer customarily signed his legal correspondence. For less formal documents he would scrawl with a flourish the initials, RRC, in a memorable, artistic script. "Call him early before the bastard sneaks out for his regularly scheduled Monday nooner," Tell added.

Contrary to the message delivered by his remarks, Tell had the highest regard for Sycamore and the Tellonics legal staff. They had kept the company on the straight and narrow – regarding matters of law, that is – for years. Tell consulted them frequently, and he instructed his executives to do likewise: "When in doubt, don't. Call a lawyer first." Unlike many CEOs, however, he had one important restriction. The lawyers were there to give legal advice, not to run the company. Sometimes they lost sight of that restriction, but Tell was always quick with a good lawyer joke they had not heard before – the more stinging the better – to remind them of the rule, and they had come to expect it. Sycamore suspected, correctly, that some of the lawyer jokes were being fed to Tell by his own staff of attorneys.

Tell and Decker discussed how Decker should comport himself at the hearings. They decided that the best way to proceed was to be cooperative and forthright, and they expected that he would receive the same advice from lawyer Sycamore. So far as either of them knew, they had done nothing illegal, unethical, or improper, although stupid was certainly a possibility. They agreed that Decker should deliver the message that the SEANET program, assuming that was their interest, was having some problems as most large scale development programs do, but corrective action was being taken in cooperation with the Navy, and the outlook was good. If they had specific questions, he should answer them as best he could without the slightest hint of evasiveness.

Tell had one final instruction: "Don't take any shit from that goddamned Flannery. The asshole usually doesn't know what the hell he's talking about, so don't hesitate to call his bluffs and refute his innuendoes." That said, Decker found himself looking forward to this new experience.

That exchange saw them safely through the salad course, and after the entrée course had been served, Tell changed the subject. "Now what's going on with Crenshaw?"

"Not sure, Arnie," Decker replied. "He's definitely not a straight shooter, but that's just an opinion. I doubt if we can prove it. Nevertheless, we may have enough hard facts to show that he has a conflict of interest and should not be serving on the Tellonics board. I expect that we'll get more information about him – or at least about his company – when I meet with Harold Laudermilk tomorrow night."

"Wake me up. I don't care what time it is. Find me and tell me what he had to say. If you can't find me, call your new sweetheart and she can find me."

Decker was shocked beyond belief. They had been so careful, or so he thought, and yet Tell knew that he and Susan Anders had been spending time together.

"Don't act so shocked," Tell said. "I suppose you think we were just driving you around in expensive limousines for the fun of it. We've been all over your ass ever since the bomb thing, and we will be for some time to come. Don't ask how."

Decker glanced at Mike Harrison. Harrison just shrugged his shoulders and smiled. He knew, so Arnie Tell knew.

"Not to worry," Tell continued. "I couldn't be more pleased. I love you both, but forget I said that when bonus time comes around. Just play it cool. So far as I know just the four of us are in on it, and as long as we keep it that way, things will be fine."

"Does Sue know you know?" Decker asked.

"Not unless you tell her," Tell replied. "I won't mention it even if you do unless you guys go public. And Mike, here, is probably more trustworthy than I am. At least he doesn't drink on the job."

Decker pondered if he should tell Susan. He decided not to decide until he had some time to think about it, but then decided to just be up front and tell her when he called later tonight.

Returning to the subject of Crenshaw, Tell continued, "Well, let me tell you that I've got some plans for Mister Mason Crenshaw that will surprise him about ten times as much as you were just surprised. I'm sure I can trust you guys, but you don't really need to know, and you're definitely better off not knowing in case it doesn't work out. But anything you learn about him from Laudermilk will be very welcome." To complete the topic, Tell instructed Decker to be very circumspect with Crenshaw. He should not let on to Crenshaw that he was suspected of at least duplicity, nor that his days were numbered as a member of the Tellonics board and candidate for CEO. At least, that was the plan.

    The three of them declined desert, but Tell and Decker had a brandy and coffee, Harrison a non-alcoholic cappuccino to conclude the dinner. Decker dutifully paid the check as agreed with his American Express Platinum card and added a twenty five percent tip (the service at Mirabelle's was first-rate as usual). They all knew that he would, of course, put the full amount on his expense report in compliance with company procedures for travel expenses not recoverable from the Government. It was not the money after all, it was the principle of the thing that was important. It was Decker's turn to pay.

After checking in at the Marriott, Decker called Susan Anders as her coded message had requested. When she answered, he said, "Arnie knows."

    "Arnie knows what?" Susan inquired.

    "About us," Jack replied.

    "Ohmygod! How did he find out? And what did he have to say about it?" Susan asked these questions more urgently.

    "It turns out that I've been under surveillance by Arnie's Army since the bomb, so they saw you at the townhouse on Sunday. I should have known, dammit. But he took it very well, said he was pleased, loved us both, and would keep it just between us – and Mike Harrison, of course, who was just doing his job --until you and I decide otherwise. He doesn't know that I'm telling you, so he won't let on that he knows even to you." Decker hoped his words would ease Susan's anxiety.

    "Well, it's probably better this way than through the rumor mill that would have picked up on it eventually. You know,

269

like that trip to the super market. It would have come out anyway," Susan reflected.

The rest of the call was a "bon voyage" from Susan and Jack's promise to call every night. Decker could not help but reflect that throughout their marriage, he had called Marty every day without exception wherever his travels took him. It was appropriate to renew the practice with the one woman whom he had taken seriously since Marty's death more than a decade ago.

CHAPTER TWENTY FOUR
REVELATION

The phone rang at seven Tuesday morning just after Decker had emerged from the shower and begun to dress in prelude to his three-minute limousine trip to LAX terminal five. It was Jane Wilson, who put Admiral Sullivan on the line as soon as she and Decker had said hello.

"Mornin', Jack," the Admiral began, then, "I haven't been able to get much information about the purpose of the SASC hearing on Thursday, but I do have some suspicions. My staff briefed Blake Blaisedale, Senator Flannery's staffer, last Thursday, and the next morning I was 'invited' to appear before the committee at this Thursday morning's session. It's obvious to me from the timing that something said or left unsaid at that briefing precipitated the hearing."

Decker answered, "That's an interesting observation. Did your staff brief him on our proposed new program plan – the satellite launch schedule changes and all that?"

"No, I instructed them not to raise the subject since the new plan was not yet approved, and I'm told that the issue did not come up. Why do you ask?" Sullivan wanted to know.

Decker said, "They may know about it anyway, and by your not mentioning it – the right strategy at the time, I agree – they may have become suspicious."

"You think there's been a leak?" Sullivan asked.

"It's possible. I've been in pretty close contact with the detective in charge of investigating the bombing and shooting incidents, and he has shared a few things with me, mostly things he needed to ask me about. One action they took was to examine Riley's telephone records to see if he might have any accomplices. There were several calls to Senator Flannery's hot line over the past two months. Detective Meisner asked me if I knew why Riley would be calling a U.S. Senator, and I told him I had no idea. Now I think I do. He may have picked up on the replanning effort going on here and passed the information to the Senator's office."

"Interesting. Yes, that would have set Blaisedale off. If he suspected we were in such serious trouble that we had to replan the program but didn't tell him about it, he would have become suspicious. Assuming that's what happened, and knowing Flannery, he wants to accuse us of misleading the U.S. Senate." Sullivan reflected that he had chosen a bad strategy in not telling Blaisedale about the replanning, even if it were still a work in progress.

Decker said, "Well, there's no way to know for sure, but that's a pretty good bet. Even so, I don't think we have to be concerned. Revising a program plan after more than two years rather than following an original plan that may no longer be valid is very common in my experience. And in my view, you were not obligated to tell anyone about it until all the details had been worked out and you had finally approved it. So if that's the issue, I don't think we have a problem. Let's just tell it like it is." Decker was relieved that he had at least some idea of the hearing's purpose.

"OK, then, that's what we'll do. In fact, I am always asked for an opening statement when I testify, so maybe I'll just lay it out for them before they ask. What do you think?" In just over a month, Sullivan had come to respect Decker's judgment.

"I agree. I may have a few things to say up front depending on the circumstances at the time, but I won't have a prepared statement." Decker wanted his options left open, he explained. If he had a prepared statement, it would become a part of the official record and might not be a good match for evolving circumstances. After years of experience in doing so, he trusted his own ability to ad lib in real time.

Sullivan thought that an excellent methodology, and the conversation ended with Sullivan's promise to keep Decker informed if any new information became available.

After his arrival at the American Airlines Admiral's Club, checking in with the receptionist and getting a cup of coffee, Decker found a quiet, unoccupied nook with a telephone on the table. Having decided not to use his cell phone for sensitive calls, the random selection of an airport telephone seemed a good choice for privacy. He talked briefly with Susan Anders and then Arnold Tell to inform him about his conversation with Admiral Sullivan. At last night's dinner he had neglected to inform Tell about Riley's calls to Senator Flannery's hot line. Learning that, Tell pondered what else Riley might have reported.

The next call was to lawyer Ray Sycamore. Sycamore was expecting the call, having been alerted by Arnie Tell just minutes before. After hearing Decker's brief recounting of events to date, he offered his advice. "Well, I think you and the Admiral have it about right. However, let me caution you about testifying under oath, which they most certainly will require. Say as little as possible in your testimony, and remember that it is testimony, not conversation. Everything you say will be 'taken down and may be used against you in a court of law' as the Scotland Yard inspector would say. Answer questions when asked truthfully, but don't offer more than is required by the question. Instead, make them ask follow up questions rather than offering information not specifically requested. If you think you should volunteer information off the cuff or feel the urge to make a speech, don't do it. In the heat of the witness chair facing two dozen U.S. Senators, you might inadvertently say something you could regret later. Your obligation is to answer their questions, not to anticipate them."

"What if they get personal and start making unwarranted accusations?" Decker expected that Flannery would do just that.

"You have a right to defend yourself, but don't appear to be defensive. I know that's hard advice to follow," Sycamore admitted, "but it is very important that you come across as frank, decisive, and believable with nothing to hide. And since that is precisely the case, you should do well."

The conversation came to a close with Sycamore's question, "Would you like to have one of the attorneys with you when you testify?"

"I don't think so. It would send the wrong message. If I have a lawyer with me it would appear that I have something to hide." Decker felt secure without real time legal assistance.

"I agree," Sycamore added. "Just wanted to make the offer. Well, good luck, and please don't hesitate to call if there are any changes in the situation."

++++++++++++++++++

The flight to Boston's Logan Airport, the short limousine drive to the Airport Hilton to check in, and the ten-minute drive to Quincy Market were uneventful. Fortunately the funneling of ten lanes of traffic into the two lanes of the Sumner tunnel was not at its worst, requiring only that the limousine driver proceed slowly, steadily, and cautiously without making eye contact with drivers in competing lanes. Decker had arranged to meet Laudermilk at the Union Oyster House, said to be the oldest operating restaurant in the country. Decker had not eaten authentic New England clam chowder and fresh caught lobster in quite some time, and he was looking forward to it.

Decker arrived first and requested a relatively isolated table where conversation was less likely to be overheard. There was always the chance that they would be seen together by someone who knew one or both of them, but Decker had calculated that on a Tuesday night Union Oyster House would be populated primarily by tourists. Laudermilk arrived shortly after

Decker and was led to the table. After they shook hands and were seated, the first order of business was to determine if Glenliven single malt was available. Indeed it was, and Decker ordered it on the rocks. Laudermilk had a draft Sam Adams dark ale with the intent of stretching out the consumption of alcohol over time since he would be driving home. He would remain committed to that delightful beverage, consuming only three pints in the two hours with Decker, well within his capacity at almost two hundred pounds.

When finally the conversation came around to the obvious point of Laudermilk's requesting their meeting, it was Laudermilk who spoke first. Decker preferred that he volunteer the information, if possible, without being prodded to do so. "I think you should know that Mason Crenshaw and Commware violated your SEANET Proposal Teaming Agreement, and have intentionally slowed progress on their part of the program."

As Admiral Sullivan would have put it, that was enough to clear his sinuses. Decker was generally aware that Commware had been a partner in developing Tellonics' proposal for the SEANET development contract, but was not aware specifically of the existence of a teaming agreement. In the pursuit of very large Government contracts, contractors tended to form themselves into teams in order to combine their resources and capabilities to their competitive advantage. In such cases, formal Teaming Agreements were executed by the parties to spell out how each would proceed in the development of the proposal and, if successful, participate in the resulting contracts. These agreements were legally binding on the parties. Usually, but not always, the parties agreed to team with each other exclusively for the limited purpose of the procurement at hand. If the agreement was exclusive, neither party was allowed to offer its services to a competing team or team member. This fact prompted Decker's question, "Was it an exclusive agreement?"

"Yes. Commware agreed to propose the development of SEANET software to Tellonics exclusively." Laudermilk's answer was offered without equivocation. He had been a participant in the Commware proposal and knew the facts.

"And how did they violate that agreement?" was Decker's next question.

Laudermilk began his extended answer. "It's complicated, but let me explain briefly what was done, then I'll answer any question you might have.

"Mason Crenshaw, Blake Blaisedale, and Ralph Lowry were fraternity brothers at Harvard. Crenshaw was the entrepreneur who started Commware in Framingham before it went public, Blaisedale was a prosecuting attorney for a time in the Middlesex country DA's office before he went to work for Senator Flannery, and Lowry, of course, became the CEO of New England Electronics. They are not only fraternity brothers, they are New Englanders, Bostonians at that, so they have very common interests and have stayed in close contact over the years. When the SEANET proposal teams were being put together, I believe that the three of them, at least Crenshaw and Lowry if not Blaisedale, decided on a strategy that would help New England Electronics. Commware would team with Tellonics and would inflate its cost proposal to drive Tellonics' cost up. At the same time, Crenshaw would secretly submit Commware's realistic estimates as a "ghost" proposal to New England Electronics. New England Electronics would incorporate that ghost data directly into its proposal as though it were going to develop the software in house. Then, when the contract was won, New England Electronics would solicit a proposal from Commware with the understanding that it would submit the same data as in the ghost proposal. They could do that after Tellonics lost the contract, because Commware was then released from its Teaming Agreement and could do as it pleased. Commware had it both ways. They would get a big subcontract no matter who won.

"So two proposals were needed. The original proposal was developed using standard cost estimating procedures. Crenshaw would secretly feed that proposal to Ralph Lowry so that New England Electronics could incorporate it as though it was an in-house software estimate. The second proposal would be artificially inflated by management direction and submitted to Tellonics. But Lowry balked at the Commware figures when he received the ghost proposal. He had other cost problems in the proposal and needed a substantial reduction from Commware to offset them. In the end, they negotiated a fifteen percent reduction in Commware's ghost proposal. So Crenshaw came to a trusted few in the proposal team, of which I was one, and instructed us to prepare an alternate proposal in which estimates were reduced across the board by fifteen percent. He did not tell us why this was being done. It was Crenshaw's intention to submit that reduced cost alternate proposal to Lowry as the ghost for New England Electronics' use. Crenshaw intended to submit the original, realistic proposal, now fifteen percent higher than the alternate ghost, to Tellonics to keep their cost for this software above that of New England electronics.

"But something went wrong. We trusted proposal team members, unaware of Crenshaw's deal with Lowry, assumed that the cost reduction proposal was intended for our Tellonics team. Crenshaw's instructions were unclear on that point since he could not let on what he was up to. Crenshaw secretly slipped a copy of the alternate ghost proposal to New England Electronics, while the proposal team, unaware of those goings on, submitted that same reduced cost proposal to Tellonics. So at the end of the day, despite Crenshaw's plan, both Tellonics and New England Tellonics got the same proposal, and both had been artificially reduced by fifteen percent at management direction. Crenshaw did not discover what had happened until it was too late to fix it. I don't think he ever told Lowry what happened, because Lowry would have no way of checking."

"Can you prove this?" Decker asked.

Laudermilk answered, "Not exactly, but I have documents that point clearly to that conclusion. I have copies of both proposals and documents that show how the cost reductions were accomplished. I have minutes of meetings with Crenshaw that document his instructions to reduce proposal cost estimates across the board. I also have the transmittal letter for the reduced cost proposal to Tellonics, which as I noted, was sent to you by mistake. The rest are emails, some of which were sent to me, and some of these are emails between Crenshaw and Lowry. Somehow I must have gotten on the wrong distribution list in Crenshaw's computer. They are not definitive, but they contain damaging inferences. In fact, it may be sufficient to show only that Crenshaw and Lowry were discussing SEANET during the proposal phase, which by terms of the Tellonics Teaming Agreement Crenshaw they should not have done. The emails confirm that those discussions took place. The documents might not hold up as proof positive in a court of law, but they would certainly cast dark shadows in the of court of public opinion or in Crenshaw's boardroom."

With that, Laudermilk handed Decker four floppy disks. "It's all here," he said. The files on the large capacity Zip disks that he had made on his office computer the night before he was fired had been copied to his home computer, and from that super set of files he had copied onto the floppy disks a subset that would be of interest to Decker.

"Why are you doing this, Harold?" Decker had taken a liking to Laudermilk and did not want to see him hurt. He was playing in the big leagues now and might not be prepared for the game.

Laudermilk responded, "For one thing, my conscience has bothered me about this for a year. When I got fired on

Friday, it pushed me over the edge, and I feel better already. Beyond that, I intend to use this as evidence that I was unlawfully terminated to keep me from making trouble. So it would be to my personal advantage if I use this information first, which I will do in a matter of hours when my attorney meets with Commware's tomorrow. If you could hold off until Thursday before using any of this, I'd appreciate it."

"Fair enough," Decker agreed. "And I will keep the source anonymous as long as I can, but eventually your name will come out. Do you understand the implications of that?

"I'm not sure what you mean," Laudermilk said.

Decker explained. "If what you suspect really happened, and it very likely did, Crenshaw might try to bring you into it as a co-conspirator since you knew about this long before you reported it. I'm not lawyer, but you should ask your lawyer about that possibility. Also the data you gave me may be marked proprietary or company private or something like that. You may have signed an agreement not to release such information to a third party, or they could just claim you stole it. You may be OK, but you should check it out with your lawyer and take it into consideration when you are negotiating with Commware."

"I don't recall that I ever signed any agreement about company private or proprietary information, and I refused to sign anything when I was fired, but you have a point. So I'll be careful." Laudermilk was now less sure of the prosecution of his intended unlawful termination suit against Commware.

Decker then added, "I think I see a way for both of us to deal with this. If we use the information only for getting the upper hand in negotiating, it may not be necessary to ever go public with it. And as you say, it might not hold up in a court of law. So let's both use it as bargaining material based on just the threat of going public. There are two, possibly three, Harvard fraternity brothers out there who would be very uncomfortable with that."

Laudermilk agreed.

279

Then Decker asked, "And what did you mean when you said that Crenshaw had deliberately slowed down progress on SEANET."

"You already have a hint of that," Laudermilk explained. "I have documented my requests for additional resources for SEANET software development, and in a number of them I submitted detailed proposals on how that might be accomplished without hurting other programs. But top management's responses were always negative. I have data – and now you have it, too -- that show the assignment of the most senior and experienced software engineers to fixed price programs, because any losses in those programs would have to be absorbed completely by the company. The cost reimbursement programs like SEANET, where profits are less dependent on performance, got the leftovers. The leftovers include a lot of very good people, but not the best and not nearly enough to do the job right. This was not a one time or temporary event. This has been a persistent problem from the beginning of the program, and Commware management has never responded to it. I can't prove it was intentional, but why else would they have let the problem linger so long. It's all there on the second floppy"

****

The return limousine trip to Logan Airport was through the Callaghan Tunnel, and at that time of night traffic flowed freely on it's two eastbound lanes, side by side with the Sumner tunnel and its two westbound lanes, also clear at that hour. Decker was back in the Hilton and dressed for bed by ten o'clock. As instructed, he dialed Arnie Tell's unlisted home telephone number.

"Who the hell is calling me on my unlisted number?" That response was not unexpected.

"Jack Decker, with the Laudermilk report as ordered," Decker replied.

"Well, that's different," Tell allowed.

Decker proceeded to recount every detail of the meeting with Laudermilk.

"That's exactly what I need," Tell stated. "Don't mention any of this to Crenshaw when you meet with him tomorrow. Just see how many more lies you can get him to tell, and remember what they were. Maybe you should prepare a memo to document your recollections of the meeting while they're still fresh in your mind.

"I've been thinking about how we might proceed with all of this in a coordinated two front attack. I'm going to make an appointment to see Crenshaw somewhere, I don't care where, on Thursday before you testify at SASC. I intend to confront him on various issues, and your data adds an important ingredient to the mix. See if you can boil down the stuff most embarrassing to Crenshaw and get it to me somehow before noon my time tomorrow.

"For your part, ask Sullivan to set up a meeting between you and Blaisedale before you testify. Insist on it. Let Blaisedale know what you know, and see how he wiggles. You can be damned sure he'll get to Flannery in microseconds. If they still want you to testify after that, it should go a lot easier."

Decker considered the various ways to get a few incriminating documents to Arnie Tell, and he decided to send them as email attachments, which he could do with his laptop computer from his room. He put the first of the floppy disks into his laptop and began to examine its contents. At midnight he had almost completed reviewing the documents and selecting the ones that he would send to Arnie Tell. Since he did not have to leave for Framingham until nine o'clock, he could complete the task in the morning when he was fresh. He called the operator and placed a wake up call for seven o'clock.

As Susan Anders had said in her coded message yesterday, it had been a long day, and that was a reminder to place the most important call of the day, Jack's personal call to Susan.

CHAPTER TWENTY FIVE
PREPARATION

The final few digits of the email to Arnie Tell were being output from Jack Decker's laptop as the bedside clock radio indicated that nine hours of Wednesday's allotted twenty-four had been expended. The transmission of twenty-seven pages of text selected for Tell's potential use had begun forty-five minutes previously and was annoyingly slow over the standard telephone line. High speed Internet connections even in first-rate hotels was not then available. You would think, Decker complained, that at these exorbitant rates they could afford DSL or cable modems in the rooms. From his townhouse he could have transmitted the lengthy message in less than five minutes. The phone rang almost immediately after the transmission was complete. It was the limousine driver inquiring if Mister Decker was ready to be transported to Framingham. Decker apologized that he was running a few minutes late, and satisfied his keeper that all was well.

Decker considered the visit to Commware in Framingham to be of little consequence, just an entertainment to fill the day until his arrival in Crystal City that evening for the important business to follow on Thursday. He did look forward to the meeting with Crenshaw to, as Arnie Tell had phrased it, "See how many more lies you can get him to tell."

The Commware building in Framingham was a recently completed concrete and glass structure of ten stories on an expansive lot of several acres covered in luxurious, green lawn except for the multi-level parking garage that approached ninety percent occupancy. Business must be pretty good, Decker concluded. He had arrived ten minutes early since traffic was light approaching the Sumner Tunnel and on the Massachusetts Turnpike. An attractive young woman of fashion model quality who introduced herself as Mister Crenshaw's guest receptionist met him in the lobby. Decker observed to himself, now there's a value added employee for a subcontractor running fifteen percent over budget.

Arnie Tell had warned Decker that Mason Crenshaw's offices were elaborate. "I've got a nice office, I admit, but Crenshaw has one of the only two oval offices of which I am aware. And his is bigger."

Miss Model, as Decker nicknamed her in his thoughts, escorted him to that elaborate suite of offices on the tenth floor and introduced him to "Mister Crenshaw's executive assistant, Marjorie Mendelsohn." She would take it from there. Mendelsohn was a pleasant woman, fifty-ish, and very well dressed. She escorted him into the oval throne room, not yet honored with Crenshaw's presence, and offered him coffee and Danish. He did take the coffee, black. They chatted briefly about Decker's trip, the weather, and other pleasantries, Decker following his usual practice of being on good terms with the secretaries/assistants of people with whom he did business. Mendelsohn informed him that they would be joined by Annita Sheldon who would conduct the facilities tour. Ms. Sheldon (Mendelsohn pronounced the "mizz" so carefully as to leave in doubt Sheldon's marital status) had arranged for other key members of the SEANET staff to join them in Mister Crenshaw's executive dining room at one o'clock after his tour. Decker asked if Harold Laudermilk and Jerome Levy would be among them. Mendelsohn hesitated, and then said, "I'm not sure if Mister Laudermilk is in today, but I believe Mister Levy will be there."

Crenshaw and Sheldon came in together at the stroke of ten. "Welcome to Commware, Jack. Hope you had a pleasant trip." After they shook hands, Crenshaw continued, "I'd like you to meet Annita Sheldon."

"Good morning, Ms. Sheldon. Very pleased to meet you." Decker emphasized the "mizz" also, but subtly, having

284

observed a solitaire engagement ring of at least four karats guarding the wedding band on Sheldon's left ring finger.

"It's my pleasure," Sheldon responded as she extended her right hand. Decker had developed the habit of always waiting for the woman to extend her hand first. "And please, it's Annita. May I call you Jack?"

"I wouldn't have it any other way, Annita," Decker smiled. She was expensively dressed also, if less conservatively than Mendelsohn. Crenshaw must pay his employees very well, he concluded. She was an attractive forty something, brown hair carefully cropped and shaped but still feminine, as tall as himself, and obviously fit.

They took seats at a group of leather chairs arranged around a glass-topped coffee table. Mendelsohn poured coffee for Crenshaw and Sheldon and topped off Decker's cup. When they were settled in and pleasantries attended to, Crenshaw began the conversation.

"I have an announcement to make, Jack. Annita is the new SEANET Program Manager for Commware," he said.

Decker was not surprised, but acted surprised, and after the briefest, but obvious, pause, reacted with, "Well, congratulations, Annita. Assuming, of course that congratulations are in order. You may have a tiger by the tail, you know."

Crenshaw added, "After our meeting in California last week, we decided that a management change was in order to help get the program back on track despite Harold Laudermilk's best efforts. I'm sorry to say that when he was relieved of his SEANET duties and assigned to another program, he chose to leave the company."

Lie one, Decker noted. Then he said, "I'm sorry to hear that. I rather took a liking to Harold, but I'm sure that Annita will do well, and I look forward to working with her." Turning to Sheldon he asked, "Have you been working on SEANET long or is this a new program for you, Annita?"

"No, I'm new to the program, but I'm a quick learner. My technical background is Ada software development, and I've been in Program Management for about five years. I'm really excited about my new job, and I'm already pretty deep into the details. I understand that you're relatively new to SEANET also," she said to Decker.

"Right, just six weeks," Decker informed her.

The conversation continued along those lines for half an hour, during which time Decker added three more items to his mental list of Crenshaw lies – or at least stretches of the facts for Sheldon's benefit. She seemed competent, and he hoped that she would be successful; however, it was possible that Crenshaw was throwing a novice into the job in order to disrupt it further. He would have to do a lot of hand holding, figuratively speaking, to make sure that didn't happen.

The tour was unexciting, as expected. The environment was mostly cubicles and offices with people in them who seemed busy at their computer screens. There were a number of conference rooms with meetings in progress, not too many water cooler gatherings, relatively quiet, and the engineering areas were well equipped with current technology software development workstations. The interesting exception was the SEANET software integration laboratory. There, Commware had assembled the hardware necessary to test and integrate their software with simulations of other system elements provided to them by DSD. Decker was pleased to see that Sheldon had become familiar with the lab in just a few days and was able to brief him knowledgeably on the status of the work that was being done there. He concluded that she had spent most of her short time on SEANET in the lab because she had determined that it was the critical nucleus of future SEANET activity.

Lunch with Crenshaw and the SEANET staff in Crenshaw's private dining room was excellent and nicely served by two well-trained, neatly uniformed waitresses. Decker was not a cost nit-picker, but he could not avoid noticing the relative extravagance of the Commware culture compared to that of Tellonics in general and Defense Systems Division in particular. Both were Spartan by comparison, although he had not perceived them as such before.

He had met most of the attendees at their program review in Buena Park the previous week, and he had a very long and informative technical conversation with Jerome Levy. They were now on first name terms, and Levy held nothing back. Sheldon listened to them attentively, interjecting an occasional question or comment, and the content of her remarks gave Decker additional confidence in her. Crenshaw spent the hour conversing mostly with the others. He knew them all by their first names, they were honored to have been invited, and the conversations appeared to be both friendly and open.   What a shame it was, Decker thought, that a guy like Crenshaw with all his talent and charisma had turned out to be dishonest. His final thought on that subject was: I pity the bastard when Arnie Tell gets through with him tomorrow.

After lunch, Crenshaw, Sheldon and Decker spent a few more minutes in the oval office. After saying his goodbyes, Miss Model appeared once more to escort him to his limousine, and Decker was off to Logan Airport to catch the next available shuttle to Ronald Reagan Washington National Airport.

+++++++++++++++++

When he checked in at the Crystal City Hyatt at six thirty, Decker was given a message to call Ken Martin at DSD Buena Park, and he returned the call as soon as he reached his room. Martin was very excited: they had discovered how Patrick Riley manipulated the CADAM system to alter previously approved drawings without changing the revision level.

"I'll say one thing for the bastard. He's damned clever. The first thing he did was get Administrator access to the CADAM system. It's not clear how he did it, but somehow he got the Administrator's password. Only the CADAM Administrator could set up and manage accounts and the passwords used to control various features that the Administrator authorized for that account.

"The way he did it was to log on to the system as Administrator to create a new identity, a fictitious draftsman that he named Timothy McVey. Can you believe that? And nobody caught it. McVey must have been one of his heroes.

"Then, logged on as Riley, he copied some of his own drawing files to the McVey archives. In this CADAM system each draftsman's entire set of drawing revisions is saved in that draftsman's unique archives for reference. Sometimes a draftsman is overloaded and needs help, and he can send drawings to another draftsman to work on them. That's why the software will permit it. The flaw in the software that Riley exploited is that files can be copied to another draftsman's archives without additional approval. .

"The files Riley copied to the fictitious McVey were previous versions one revision letter below the current revision level. After that was done, he deleted the current revision of those drawings from his Riley archives. Then he logged on as McVey and revised the drawings to the next level, but put in the damaging differences. He approved them with an engineering password he had created for McVey, and they were automatically updated in McVey's archives to the next revision letter -- the same level as the drawings deleted in Riley's files. Next McVey copied the revised drawings back to Riley. While one user can copy files to another user, the software does not allow one user to overwrite existing files of another user, and that's why Riley had removed the latest revision files from his archives. So the drawings with unauthorized changes replaced the correct ones of the latest revision level that had been deleted from Riley's archives

"The CADAM system retains a fairly detailed log of file transfers, so we can trace what he did and when he did it. We can see that he did this switch methodically over a period of several weeks, and the changes were so small that they were obscure to an observer, but they were big enough to screw up critical impedance matches on the microwave boards. The important thing is that we can identify all of the drawings he changed, and fortunately there were not very many. We can recreate the correct drawings from the existing engineering data that was used previously to make the correct revisions. And the graybeards are certain that the subtle changes were sufficient to cause the problems. It explains all their test data.

"It's going to take a few days, maybe a week, before we can send the corrected CADAM to the board vendor to build new boards. The bad ones, and the components soldered on them, are now just scrap – very expensive scrap. Bob Masters is ordering replacement parts on a crash basis, and Ralph Brown thinks he can start delivering good boards in about six weeks. It fits, but just barely. We'll have to watch it very closely."

Decker was more relieved than he could express to Martin, but he tried. "Ken, that's the best news I've had since I came back to DSD. I can't tell you how relieved I am. Just in time, too. I'm sure Admiral Sullivan will be just as pleased as I am, and if it should happen to come up during my testimony this afternoon, you may have saved my ass. Good job, Ken, and please pass along my thanks to all the people who worked on it. I'll see you all on Friday, and we'll have a celebration."

****

Admiral Sullivan was in his office when Decker arrived at seven thirty. The Air Force General would arrive at eight for their

289

meeting, and Sullivan invited Decker to wait in his office. Sullivan admitted that he had not been able to learn any more about the reason for the hurriedly called SASC hearings, and they confirmed their strategy on the assumption that Senator Flannery would raise the replanning issue, They wondered what else Flannery might have in store for them, but at that moment they had no clue.

Decker shared the good news about the microwave board problem with the Admiral, and also told him that their experiments with the faster microprocessors from Zybyte were very encouraging. Milt Karinski was certain that a combination of fast processors and Ada software simplification would bring the Satellite software timing problem under control. DSD was still working on the Ada to C++ translator just in case they did not. The Admiral could honestly report to SASC that there were no known technical showstoppers for which reasonable solutions were not available.

Then Decker asked the Admiral to set up a meeting for him with Blake Blaisedale before he testified in the afternoon, and the Admiral asked Jane Wilson to get Blaisedale on the phone. Sullivan arranged for Decker to meet Blaisedale in the Capitol Building Rotunda at one o'clock, and they would find a quiet place to talk for a few minutes before the hearings resumed with Decker's scheduled testimony.

When asked why he wanted to meet with Blaisedale, Decker replied simply, "It's just contractor stuff. You wouldn't be interested." Decker wanted Admiral Sullivan insulated from any fallout that might result from the industrial strength bombs he might be forced to drop on Blaisedale and Flannery. Trusting Decker that it was in his best interest not to know, the Admiral dropped the subject.

General Scott arrived promptly at eight o'clock, accompanied by Major Randy Garner, the SEANET project officer who had represented the Pacific Missile Test Center during the Navy's program review. Admiral Sullivan knew them both, and Decker was surprised to find that he did, too – Garner from the recent program review, and Scott from over fifteen years ago.

As they shook hands and exchanged greetings, the General said to Decker, "Remember me, Jack?"

"Of course I remember you, 'Major' Scott." Decker turned to Sullivan and explained, "The General and I, at that time the Major, worked together on the Northwind project back in the eighties." Then to the General he said, "I had no idea that you were the same Joe Scott."

"And I had no idea you were the same Jack Decker. This is a very pleasant surprise. You, Admiral, are very fortunate to have this guy on your program. We had some very exciting days on Northwind, and it was a great success --not without considerable credit to Jack, I might add."

What a great stroke of luck, Sullivan thought. These guys not only know each other, they appear to have mutual respect for one another. It couldn't hurt. "That's good to know, General. My apprehensions about Mister Decker's capabilities are eased somewhat by your vote of confidence." The Admiral's tongue in cheek comment was understood by all present to be a compliment to Decker. "I know you have a tight schedule, Joe, so shall we get right down to the issues we have to resolve?"

They all took seats at the Admiral's conference table, and Jane Wilson offered them coffee, tea, or soda, and they all had coffee. General Scott then brought them to the heart of the discussion. "Actually, Rusty and Jack, there is only one issue. The good Major has convinced me that you guys know what you are doing and that a change in the satellite launch schedule is the best way for your program to proceed. We don't resist schedule changes, despite what you may believe, just to be difficult -- although some of us tend to be a little ornery on the subject. We resist because it's expensive. It takes a lot of detailed planning and coordination to launch a satellite. Things have to line up just right. So if we change a schedule, a lot of that planning and coordination has to be done over again, and it's expensive. My problem is that your schedule changes for three satellite launches are going to cost PMR about a half million dollars more than you have budgeted for us, and that's a half million dollars I don't have. You'll have to pay the piper, Rusty."

Sullivan's answer was, "Well, if that's the case, then we do have a problem. I can cover all the bets we have on the table at the moment, but I don't have any loose change lying around – and certainly not a half million dollars."

"I think we can help," Decker interjected. He was about to make a commitment for the corporation somewhat beyond his authority, but he believed that Arnie Tell would agree that it was the right thing to do. Decker knew that without the changes, Tellonics stood to loose much more than half a million. There would be serious delays and additional overrun without the revised schedule. It was a reasonable, if unanticipated, tradeoff.

"Tellonics will cover your added cost, General. Give me a couple of days to work it out back home, and I'll come up with it somehow."

"Now there's a switch," Scott said. "The contractor is going to pay the Air Force. How 'bout that?"

"Cost reimbursement only, Joe. Your actual cost and no fee. I'll send my auditors over to watch your every step and count every penny," Decker replied. They all laughed at that reversal of roles.

The discussion continued for another ten minutes even though the central issue had been resolved. Decker would, by some legitimate process and on a schedule to be determined, reduce Tellonics' billing to the Government by five hundred thousand dollars, and the Navy would increase PMR's funding for SEANET support by that same amount. At this point it was only a gentlemen's agreement, but the three of them considered it a done deal.   The DSD, Navy, and Air Force   contract administrators would have to work out the details.

The meeting ended with General Scott's sincere invitation to both Sullivan and Decker to visit his headquarters in the near future and to observe the SEANET satellite launches when they occurred. He made it clear that they would be the personal guests of the Commander of the Pacific Missile Range.

Just as this early morning meeting in Crystal City came to a close, Arnold Tell's limousine pulled into the entrance of Commware in Framingham.

CHAPTER TWENTY SIX
ONE ON ONE

Arnold Tell called Mason Crenshaw directly from home at six o'clock Wednesday morning, nine in the East, to arrange a Thursday morning meeting. Crenshaw was reluctant at first, complaining that he was booked solid that day, but Tell insisted and offered to do the traveling so that Crenshaw would not have to leave Framingham. When asked what the subject was and why the urgency, Tell replied, "We have some imperative corporate business that can't be discussed on the telephone, and it has to be taken care of this week. We have to talk one on one. I'll see you at your office at nine o'clock tomorrow." Reluctantly, Crenshaw agreed.

When he reached his own office Tell asked Susan Anders to arrange for the Cessna Sovereign to depart Orange County's John Wayne airport at noon for a trip to Worcester, Massachusetts, returning Thursday at noon. By landing at Worcester he would avoid Logan Airport and the dreaded Sumner and Callahan Tunnels. In addition he would need a hotel reservation tonight and ground transportation to Commware's offices for a nine o'clock meeting Thursday. Anders suggested he stay in Worcester since he would be arriving late, and the morning trip to Framingham would take less than half an hour. Tell also alerted her that Jack Decker would be transmitting a long message to him this morning, he was not sure how, and he asked her to be on the lookout for it.

Anders had access to Tell's email and filtered it for him in much the same way as Jack Decker had arranged with Erik Hansen.

"It's already here," she told him. "The world's longest email. I printed it out for you, but you can still see it on your computer. Not to eavesdrop, but I couldn't resist scanning a page or two. Unbelievable."

"Well, if it's what I think it is, you're right. Unbelievable, and just in the nick of time." Tell could hardly wait to begin reading the material Decker had sent him and to think through how he would use it in the Thursday meeting with Crenshaw.

.   Of the three Tellonics executive jets, which also included a Lear 55C/ER and a Hawker 800XP, Tell preferred the Cessna Sovereign for non-stop coast to coast trips, although he often joked that he could "make do" with the Lear or the Hawker. All three were fine aircraft, each with its own personality, and As CEO he had first priority on their use. The executives reporting directly to him could use them for business travel, and the cost was considered a corporate level general and administrative expense for tax purposes. Executives and managers from the operating divisions could use them if they were available provided that they paid a mileage assessment. It was economic for the divisions to use them when as many as five people were aboard since the allocated mileage cost was usually less than the cost of five individual commercial airline tickets. Often a further saving could be achieved by avoiding overnight stays since the aircraft could use small airports near destinations and could come and go at any time, air traffic control and weather permitting. Best of all, no charge was assessed if division passengers could "catch a ride" with a corporate executive who happened to have the same itinerary and agreed.

Tell's trip proceeded as planned. The Sovereign lifted off from John Wayne at twelve eighteen and touched down at Worcester at eight twenty nine local, just over five hours. Since the hour was late, Tell invited the crew, pilot and copilot, to share his limousine to the hotel, and they had dinner together also. Naturally the talk was about aviation, especially navigation, the latter a major product line of the Tellonics Avionics Division in Tulsa, Oklahoma.

The Thursday morning limousine deposited Tell at the Commware offices just before nine o'clock, and he received the same welcome as Jack Decker had the day before. He, too, thought Miss Model a bit of an extravagance, a little thin for his taste, but he jokingly hit on her anyway just to stay in practice. She was politely unreceptive.

Although Tellonics was a major corporation of five billion dollars annual sales and Commware one tenth the size, both Tell and Crenshaw wore the mantle of Chief Executive Officer, and both were accustomed to considerable deference, especially when on their home turf. The difference was that Tell put up with it and Crenshaw demanded it. Tell would have preferred to have this meeting in Burbank where he was in charge. Nevertheless, he reasoned that even though Crenshaw would have home field advantage, he would also be on the defensive --advantage Tell. Often when he needed to speak with a staff member, Tell would just wander into that person's office rather than call the person to his own office. In that way he could control both the agenda and the length of the meeting. When his agenda was accomplished, he could just leave. The same rule applied here.

As the host Crenshaw welcomed Tell cordially but reminded him that he had only one hour set aside for their meeting. He was scheduled to meet with the Commware board of directors at eleven and needed to spend at least an hour with members of his staff to prepare for that meeting.

Rather abruptly Crenshaw asked, "What is the imperative corporate business that we couldn't talk about on the phone?"

Tell had resolved to keep his temper and vocabulary under reasonable control and to be as civil as possible despite the need to confront Crenshaw. "There are two things, Mason. Neither of them will be pleasant for you, and I am very sorry that it has come to this. However, you leave me no choice.

296

"Let's take the easier one first. We are both aware – please don't deny it – that you want my job as CEO of Tellonics. That's OK. You are free to pursue that goal so long as you do it lawfully and ethically.   The SEC won't stand for anything unlawful, and the honorable members of the Tellonics board won't put up with any funny business.   Under those same constraints, I am entitled to defend my position, which I intend to do, and vigorously. You've been campaigning for proxy votes so that you can control the composition of the Tellonics board and have me ousted, but I have not been idly watching. Ever since you started your proxy campaign, and with all legal requirements met, I have been buying outstanding Tellonics shares on the open market through my broker, and at this point I own personally almost forty percent of Tellonics' outstanding shares. All I need to defeat you is another eleven percent. Over your objections for the past several years the board has authorized the distribution of stock to our senior executives as incentives, and at this point they own another fifteen percent.   Of those shares I have commitments for their proxies to give me an additional eleven percent and counting.   In other words, I will control over fifty percent of the voting stock of the company, and you are wasting your time."

Crenshaw's response was as Tell had expected.   "I'm not stupid, Arnie. I know what you've been doing, but don't be so naïve as to think I can't change the minds of some of your VPs. Just remember that we all look out for number one, and when the chips are down, I don't think you will keep enough of their votes to win. You see, you and I have a different point of view. I will use profits from current operations to put dividend money in the pockets of shareholders – including your VPs. And as a result the value of their shares – and yours – will rise. You, on the other hand, want to keep most of the profits for the company in order to expand. Well, I say Tellonics is big enough. You may have verbal commitments from your executives, Arnie, but I will have their proxies before the next board meeting. It's time for the big payday."

"You could be right, Mason. Except for one little thing," Tell said.

"What's that," Crenshaw asked.

"You're going to jail," Tell replied.

"What the hell are you talking about?" Crenshaw asked, astonished.

"I have here documentary evidence, which I will leave with you to ponder, that shows clearly that you have broken the law. Tell reached for his briefcase, extracted a bundle of documents bound with rubber bands, and placed them on the table in front of Crenshaw.

"Your last two annual reports and your quarterly earnings reports grossly exaggerate the value of your actual contracts and the revenue they produce. I don't understand how your auditors missed it, unless they were in on it, but our accountants just happened to notice the overstatement of your SEANET contract, which happens to be your biggest. That got our attention, so we had our outside auditors quietly look into some of your other contracts, and they found more deception in your public disclosures. You've been cooking the books, Mason, ever since you went public to attract investment so you could build this extravagant palace. I intend to bring this to the Government's attention. I expect that by the time the SEC and the Justice Department are through with you, you'll be doing hard time, and when you get out you'll be barred from serving on any board of directors under SEC cognizance for the rest of your life."

"You're full of shit, Tell," Crenshaw yelled.

Tell said quietly, "Check out the documents, Mason, then we'll see who's full of shit."

"Furthermore," Tell continued, "Tellonics is going to acquire Commware. In other words, I'm buying your stupid little company, and the next thing I'm going to do after I fire your ass is to get rid of that "guest receptionist" you've been fucking."

"You don't have the money to buy Commware," Crenshaw asserted without denying the allegation regarding Miss Model.

"I don't need money. I've got Tellonics stock worth ten times the value of your little operation here even with your illegally overstated assets. Tomorrow afternoon just after the stock exchange closes my bankers will be offering to exchange a share of Tellonics stock for every two shares of Commware. That's thirty percent more than market value. And I don't need the board's approval to sell my personal stock. I have filed my intentions with the SEC, and the only other thing I have to do is give thirty days public notice before I actually complete the transaction. You and your shareholders will have to wait a month to get your thirty percent profit on your holdings. Just how many of them do you think will turn down a deal like that? You'll make a bundle yourself, Mason, so don't complain too much. You can use it to bribe the guards and buy pot to smoke and coke to sniff in your cell."

Crenshaw was finally beginning to realize that Tell was serious and might not be bluffing.

"Oh, the other thing. I almost forgot. I'm bringing suit against Commware for a hundred million dollars in damages resulting from your breach of our SEANET teaming agreement." Tell extracted the second bundle of documents from his briefcase and put them on the table next to the first bundle. These were copies of the Laudermilk papers. "You don't have to read all of these since you wrote most of them -- including the ones that show how you deliberately withheld available resources from SEANET in favor of your fixed price contracts. That could be worth some hard time, too, but maybe the judge will go easy on you and let your serve the sentences concurrently."

"How did you get these? These are company private documents. You could go to jail, too, you know, for stealing them." Crenshaw had his first rebuttal opportunity.

Tell was unyielding in his attack. "Well, I didn't steal them. They were provided to me by your former employees who maybe stole them and maybe not. If I go to jail, I'll try to get the cell next to yours so we can play checkers through the bars."

Finally shaken, Crenshaw asked, "So what happens next?"

Tell made his proposition. "Tell you what, Mason. I'm a nice guy, and I'm willing to forgive and forget most of this stuff – on certain conditions, that is.

"First, on your half-assed Tellonics board takeover attempt, back off and I'll keep that first bundle of papers to myself. Then, when I get your resignation from the Tellonics board, which I will publicly regret since you're such a fine fellow, I'll burn them and all existing copies. Well, all but the originals, which I will retain for a rainy day.

"I'll probably drop the lawsuit for breaching the Teaming Agreement since I'd just be suing myself. I do intend to buy Commware so that I don't have to put up with your bullshit any longer. That second stack of papers shows that you and your lousy fraternity brother at New England Electronics broke enough rules to get both companies disbarred from doing Government business. That should go over big with Alpha Beta Delta fraternity or whatever the fuck it's called.

"This is probably not one of your best days, Mason, but you do have one bit of luck. I've got a problem you can help me with. My guy Jack Decker is scheduled to testify at SASC this afternoon in front of that asshole Chairman, your very own Senator Flannery. His main man Blaisedale is also a frat bro of yours, so you just pick up that telephone when I leave and tell him to get his boss to lay off Decker or I'm dropping a copy of this stuff on the Senator's colleagues so they can have a real hearing. I expect you may be the guest of honor.

"So it ain't all that bad, Mason. You don't have to show up at the Tellonics board meetings anymore, you exchange your Commware shares for Tellonics shares and make a nice profit, and you don't get any subpoenas from the U-S-of-A Senate. What's more you'd better keep this to yourself until it's announced. Any hint of insider trading before my offer is made public tomorrow and I'll turn the lot of you in, and the deal is off. Instead, once announced, I insist that you get your board to endorse the sale before the exchange opens on Monday.

"I'd say that's a fair arrangement on balance, so how 'bout it, Mason?"

"I'll get back to you after I talk with my counsel," Crenshaw responded.

"You don't need a goddam lawyer to tell you that you broke the law and you're going to jail. We both know it, and so will your fucking counsel. No, Mason, this is a one-time offer. I'll give you five minutes to leaf through those papers, and then I'll just have to have your answer. Your call, Mason. Take the offer or take the consequences."

"I don't know if Blake can get Senator Flannery to back off," Crenshaw admitted, clearly wavering in favor of Tell's offer.

"Well, by God, you'd better hope he can. Maybe your whole fucking fraternity should call Joe Flannery and promise to steer all their PAC money to his next campaign – in addition to the under the table cash they've been sending him. Tell the Honorable Senator Flannery I've got a couple of papers on that subject, too." This time Tell was bluffing, but on a hunch he did not expect to be called. He was not.

Crenshaw looked at a few of the papers. He was not certain of the implications, but he was convinced that Tell would not have provided him these copies if they were not sufficient to support his allegations – allegations that Crenshaw knew to be true. He decided to cave. Tomorrow was another day, and he would have considerable capital from the exchange of stock.

"Deal," Crenshaw said. "We'll do what we can with the Senator. And don't let the door bang you in the ass on your way out."

"You don't need to worry about my ass, Mason, you need to worry about your own. You go back on your word this time and your ass will be in a big-time sling. You can count on it." With that parting volley, Tell picked up his briefcase, turned away and left the office. He did not wait for Miss Model to escort him; he took the elevator directly to the lobby and walked to the waiting limousine.

CHAPTER TWENTY SEVEN
THE HEARING

Of the twenty-three members of the Senate Armed Services committee, only sixteen were present when the meeting was called to order at ten o'clock, by chance an equal number of Democrats and Republicans. That distribution could change at any moment as Senators came from and went to other activities according to their individual priorities. For the Chairman, Senator Joseph Flannery, this was the highest priority. He had waited for almost two years to have Rear Admiral Russell Sullivan testifying before his committee in a compromised position. It was Flannery's intent to embarrass this officer to the extent that he would have no choice but to resign in disgrace, this officer who had, in the Senator's view, unjustly awarded the SEANET contract to Tellonics when by all rights it should have gone to his home state's New England Electronics. The Admiral was seated at the witness table before him and, having been caught in a deception, would soon be squirming helplessly despite the gold braid and chest full of ribbons.

"Good morning, Admiral," Flannery began, "glad to have you with us again. I hope you will forgive the haste with which it was necessary to have you come over, but a serious matter has come to our attention regarding the SEANET program that must be treated with some urgency. I will direct the committee's attention to that matter promptly; however, as is our custom, would you please provide us a brief summary of the status of the program for which you are the guardian of the people's interests?" Flannery sat back in his chair with a broad smile in anticipation of Sullivan's delivery of a capsule summary of the briefing that been given to Blake Blaisedale just a week previously, a briefing in which it was not disclosed to the Senate's official representative the fact that the program was in such severe difficulty as to require major replanning.

The "serious matter" had not yet been identified by the Senator, but Sullivan and Decker had calculated that it was the undisclosed replanning. On that assumption, the Admiral would disclose it now, albeit somewhat prematurely, in order to defuse the Senator's expected attack. Decker and his staff had solutions to the major technical problems, a plausible schedule had been worked out, and General Scott had been persuaded to reschedule the satellite launches. If the undisclosed replanning was the issue, Sullivan would confront it head on.

Sullivan began, "Thank you Mister Chairman, and good morning to you all. Among my responsibilities as a major program manager for the Department of Defense, I regard my appearances before this committee as the most important because, as the Chairman has just pointed out, I am designated the guardian of the people's interests on a program which the Navy considers of the highest priority."

Sullivan then consulted his notes. They were just notes, bullets, talking points to guide his opening statement. A copy would be provided to the committee, but they were of little consequence without the accompanying word-for-word transcription of Sullivan's testimony that was being made in real time. Sullivan, as Decker had suggested, realized that without a literal document, he had considerable latitude available in his choice of words and emphasis.

He continued. "In naval operations nothing is more vital than the ability of our globally deployed ships, submarines, aircraft, shore stations, and command centers to communicate rapidly, reliably and securely, and we are well down the road toward providing an order of magnitude improvement in that capability with the SEANET program. The very best of Government and industry has worked to solve this very difficult problem for more than ten years, and today we are approaching the end of the engineering development phase of our effort. Based on my understanding of our current status, I can report with confidence that in less than one year we will have demonstrated that much needed capability through the efforts of the Navy's Operational Test and Evaluation Force. That accomplished, we will be ready to embark on the initial production phase of the program, and in a very few years we will have deployed that capability throughout the Navy.

"A great deal of hard work and ingenuity has gone into the program to date, and we have spent and will spend a great deal of the people's treasure, which you have allocated to us, and of which I am the guardian. I am well pleased with what has been accomplished so far. But in all candor, I must also report to you that what lies ahead in the coming months will be just as difficult as what has gone before. In anticipating that difficulty several months ago, I asked our prime contractor, Tellonics, to reexamine the program in all its dimensions. They have done so, they have made some specific recommendations, and we have just this week completed our review of their recommendations.

"One of the first things that our prime contractor did, and possibly the most important, was to bring onto the program the most accomplished program manager in its long history dating back to World War one. I understand that he, Mister Jack Decker, will be testifying before you this afternoon, and I think I will just find a seat in the audience, if one is available for me, to hear what promises to be a most enlightening exchange.

"Under Mister Decker's guidance, every detail of the work remaining to be done has been examined, reexamined, and reexamined again. It's important for you to know that the Navy has been a full participant in that reexamination, and we have concluded several things. There are three or possibly four severe technical problems that need resolution in the near term. Those problems are well understood, solutions have been proposed, and they are being implemented and verified. You may find it interesting that I have learned a few lessons about practical engineering from Jack Decker. At the top of the list is this one: first, take the time to understand the problem completely and unambiguously; then, and only then, can it be solved. That lesson can be applied as well to many things beyond engineering, but that's a subject for another day. Of course, in a program of this magnitude and complexity, there are a number of less critical engineering problems facing us, but we see nothing there that will require us to violate the laws of physics or exceed our resources. I am quite confident that we will solve these routine problems in due course without major impact.

"We are also concerned, as I know you are, about the cost of the program. At this moment we have spent about ten percent more than we had estimated to bring us to this point. In the arena of high technology development programs, a ten percent overrun, while not desirable, is considerably less than average. Prudently anticipating the eventuality of an overrun, the Navy set aside sufficient reserves to cover it, so we do not expect to come back to you for additional funding. What's more, we are determined to complete the remainder of the program for a bit less than originally estimated. Now, that must sound overly optimistic to you, but as a matter of fact our navy-industry team has found ways to complete the program on time but at less cost than originally estimated for the remaining work.

"I hope you will forgive me for not disclosing our new plan until we had completed it and I had approved it. I concluded that premature announcement of a major replanning effort would produce unwarranted concern and unintended consequences. To avoid them, it was my decision, and mine alone, to keep our planning effort to a trusted few until it was completed. That may have been a mistake politically, but that's the way I was taught to do things at the Naval Academy. Of course, as you must have noticed, I am not much of a politician, but I was a pretty good fighter pilot and nuclear aircraft carrier captain, so perhaps you will overlook my ineptitude at Pentagon politics.

"Simply stated, the plan is this. Rather than launch all three of our development phase satellites in close proximity as originally planned, we will launch them just when they are needed. By relieving the pressure created by an unnecessarily compressed satellite launch schedule, we can do other work in a more orderly and less costly manner. Another element of the plan is to reduce the amount of testing to the essentials. We identified a great many tests that were being done not because they needed to be done but because the 'textbook' or the 'specification' said to do them. We also found that, just for the sake of completeness, we were attempting to test requirements that were not reasonably testable. For example, we had included an exhaustive and very expensive test to verify by spectral analysis that we were using exactly the right color of paint on various antennas. We decided that we could do that well enough with our eyes and a color chart. And many tests were to be repeated again and again at successive levels of integration. So when we could reasonably eliminate tests without adding risk, we have taken the redundancies out of the test program, unneeded tests that would be expensive in terms of both time and money. If you have concerns about the advisability of these simplifications, please remember that OPTEVFOR is going to give the system a damned thorough workout before they release it for operational use. That's your insurance policy.

"Finally, although there is considerably more to it, we have changed or eliminated some unrealistic and unimportant specification requirements because of another lesson I learned lately. I hesitate to credit Mister Decker again lest you think I agree with his every suggestion, which I certainly do not – we have had and will have some heated exchanges -- but he did bring this fact to my attention. Just because the Navy writes a specification requirement does not mean that it can actually be physically accomplished within the constraints of time and money available to us. Furthermore, many requirements can be classified as "bells and whistles" whose expense does not justify their incorporation. The Navy is taking care of those excesses in the specifications without impacting the essential performance of the system.

.   "Taken together, I am convinced that these measures will see us through to a successful completion. We have a lot of hard work to do, but we have the resources and the will to get it done.

"In conclusion, I invite you to have your technical and administrative staff members examine our new plan in all its details. If you have questions, and by all means good suggestions, please let us hear them.

`"Thank you Mister chairman, and now I will be pleased to answer any questions the Committee may have." Sullivan's remarks had been right on target, and the Committee had been attentive to his every word, including those Senators who entered the chamber late.

Flannery was no longer smiling. He had been preempted on his main point of contention with the Admiral, and for the moment he was forced to retreat to lesser controversies. Besides, there was still Decker in the afternoon.

After a few moments of silence in the room, Flannery spoke. "Thank you for your remarks, Admiral, eloquently phrased and informative as usual. I'm certain that my colleagues will agree that you were, indeed, much more than a 'pretty good' fighter pilot and nuclear aircraft carrier captain. As for your political prowess, let me just state that I sincerely hope you do not intend to run for United States Senator from the Commonwealth of Massachusetts." The audience, Senators, Sullivan, and spectators of every stripe erupted in good-natured laughter. That sound byte would certainly make the evening news on all networks. In fact, rumors had spread throughout the Capitol building and into the pressroom that something interesting was going on in the SASC hearing room. CSPAN had it all on videotape for the media pool to excerpt, and there was more to come.

The questioning of the Admiral was hardly controversial and of only fifteen minutes duration. The committee members were anxiously waiting for Flannery to raise his "urgent matter," but it was not forthcoming. Out of respect for the Chairman, he was not pressed on the issue. When questioning came to an end, Flannery adjourned the meeting until two o'clock, at which time "... the highly praised Mister Jack Decker ..." as he put it, would be testifying. The press, which had gathered in larger numbers now, was mystified. Perhaps the "urgent matter" would be disclosed in the afternoon session during the testimony of Jack Decker.

+++++++++++++++++++++

Decker's limousine deposited him at the steps of the Capitol Building at five minutes 'till one. He walked up the broad steps briskly, and after being detained for a security check, he walked into the expansive rotunda. Even though well into September, and a Thursday at that, the area was bustling with tourists, but not overcrowded as it sometimes was. He spotted Blaisedale, or the man he presumed to be Blaisedale, at once standing near the bronze bust of Abraham Lincoln as Admiral Sullivan had arranged. He walked over to the man and said, "Mr. Blaisedale, I presume?"

"Yes, I'm Blake. You must be Jack Decker," he responded. They shook hands, and Blaisedale continued. "Let's walk down the hall to a small room I reserved so we can talk privately."

"Thanks," said Decker. "That would be best."

The small room, like the rest of the Capitol Building, was decorated with portraits of former legislators, these somewhat lesser known personalities that Decker did not recognize. There was a small round table and two chairs, period pieces in cherry that would be worth a small fortune on the antique market. "Have a seat," Blaisedale said. "Sorry I couldn't arrange for refreshments on short notice."

"Not a problem," Decker said. "I've just had too much lunch, as I always do when traveling." Blaisedale agreed that it was a universal problem. Then Decker got to the real purpose of the meeting. "I do want to thank you for seeing me. You must be terribly busy, so I won't take much of your time, but I believe it was important to have a word with you before my testimony this afternoon."

"Well, as you can imagine, I can't really discuss your testimony with you. That is between you and the committee members. I'm just an advisor," Blaisedale said. "If there is something other than the specifics of your testimony that you want me to discuss, that will be just fine."

"It's a little hard to separate the wheat from the chaff in this case," Decker said. "Especially since I don't even know why I'm here. But I suspect it must be related to the SEANET program. No discussion is necessary. I just want to provide some information that you may want to pass along to Senator Flannery."

"I can tell you why you were called. The Senators have some concerns about SEANET that they would like to hear about directly from you. So what is it you need to tell me, Jack?" Blaisedale felt comfortable with shifting to first names even though he had just met Decker. Perhaps the Admiral's remarks about him in the morning session had persuaded him that Decker must be trustworthy.

Decker had taken the measure of Blaisedale also, but with less to go on. All he really knew was that he had been in public service for most of his career, first as an assistant DA and then as a member of Senator Flannery's staff. He was a gifted lawyer, so he could have done much better financially outside of Government. Decker decided to give him the benefit of the doubt by assuming that he was not a party to his fraternity mates' conspiracy unless he received contrary indication. "You, and Senator Flannery, need to know that we have just discovered an unethical, and possibly illegal, action on the part of two of your personal friends who have been involved in the SEANET evolution: Mason Crenshaw and Ralph Lowry."

"You're kidding." Blaisedale's disbelief seemed genuine.

"I'm afraid not. Crenshaw violated the terms of our exclusive teaming agreement and passed protected information to New England Electronics during the SEANET competition – information that found its way into New England's proposal. We have documentary evidence, and we have the word of a prior high-level employee of Commware. It's pretty clear what they did."

"That's a serious allegation. You say you have documentary evidence?" Blaisedale still found it hard to believe.

"Yes, we do," Decker responded. "The reason I bring this to your attention is that I have no desire to disclose it to the committee unless circumstances demand it. You may or may not believe this, but Tellonics is one of the good guys in the corporate world. If your Senator follows his usual pattern of ad hominem attacks and unsupported allegations, I will be compelled to lay this out as an example of what the bad guys are doing – and they're both Massachusetts companies as you know. You need to understand that if pressed too hard, I will testify to this."

"I'll pass your information to the Senator," was Blaisedale's only comment. No discussion was necessary. "Anything else, Jack?"

"That's about it, Blake. Thanks for seeing me, and I expect that we'll meet again – on a more pleasant occasion I hope. See you at two." Having delivered his message, Decker rose, as did Blaisedale, and they went their separate ways.

This was not the first message that Blaisedale had received that morning. He had received frantic calls from fraternity brothers Mason Crenshaw and Ralph Lowry. They had essentially the same request: don't go too hard on Tellonics. It would not be in the best interests of the Commonwealth of Massachusetts if certain allegations by Tellonics were made public at this time. Talks with Tellonics were in progress, and Decker might release information that that would be harmful to both Massachusetts companies. Blaisedale had passed the requests along to the Senator, but he did not seem particularly interested in them. Having learned from Decker the reason for their urgent requests, he expected the Senator might take the matter more seriously.

Senator Flannery had received several calls of his own following Admiral Sullivan's testimony in the morning. Two were from newspaper reporters to whom he regularly leaked information that would serve his purposes, particularly information that would reflect unfavorably on "the other side," as the opposition party is called in the Senate. He told them he would call them back at the end of the day when he had more to disclose – off the record, of course. The other important call was received just as Blake Blaisedale was having his meeting with Jack Decker. It was from retired Comptroller General Keith Majors.

"Good afternoon, Keith. How's the retirement going?" Flannery was cordial to Majors although he never cared for him very much. Majors still had high-level contacts in the Government, and therefore commanded respect.

"I'm doing rather well, Mister Chairman," Majors offered. "Retirement is highly recommended while one can still sit up and take nourishment."

The pleasantries continued for another few sentences, and then Flannery asked, "How can I be of service to an old friend today?"

Majors replied cautiously. "Well, I'm pleased that I am still your friend, Joe, despite a few differences we may have had over the years. I just wanted you to know that I caught some of your SEANET hearing on CSPAN this morning. I thought your comment on Admiral Sullivan's political prowess was a real jewel."

"Well, thank you Keith. It did go rather well I thought." Flannery was still curious why he was being called.

"I was quite impressed with what the Admiral had to say, and I understand that you're going to interview the prime contractor this afternoon. That's why I'm calling, Joe. There's something you need to know about the SEANET procurement."

"OK, Keith, let's have it." Flannery was getting impatient, but still spoke politely.

"You recall that there was a protest of the award to Tellonics by your constituent, New England Electronics. Among other things that GAO looked into to resolve that dispute were the detailed cost proposals. We found something curious – no, disturbing – but we didn't report it because it would have been too difficult to substantiate. You see, the New England Electronics' cost proposal contained a section on satellite software development and its estimated cost. Tellonics proposed to subcontract that software to another of your constituents, Commware. The Tellonics proposal contained a cost package developed by Commware for that same work. The two cost estimates were identical in important respects – not similar, but identical. I suppose that two companies could develop identical cost estimates, but for a job of that magnitude, and high-risk software development at that, it was one hell of a coincidence.

"I've been thinking about it this morning Joe, and it occurs to me that if this is something other than a coincidence, it just might come out in the wash if you start pushing the contractor around and digging into things again. That wouldn't look too good for either of us, but I'm retired now. You're not. I just wanted to pass this along for your consideration before you begin the afternoon hearing."

Flannery did not consider this damaging. "Well, thanks for the heads up, Joe, but I really don't see how that could affect me. I didn't know anything was out of line. You're the one who didn't report it."

"That's true, Joe," Majors added. "But there's a related point I wanted to make. You recall that you put quite a lot of pressure on me to make the SEANET protest come out in favor of New England Electronics. That's not ethical, Joe, and your colleagues would be very disturbed if they knew about it. It might even be illegal."

"Pressure? What pressure?" Flannery was in denial mode.

Majors fired his shot across the bow. "Perhaps a replay of some of our telephone conversations would refresh your memory, Joe. Would you like to hear one of them now?"

"You sonovabitch! You taped our phone calls? That's not legal, you know, to tape a phone call without both parties' permission." Flannery went on the offense.

Majors had the best of the argument. "So sue me, Joe, or call the Attorney General. Of course, they'll have to listen to the tapes if you report me. And what happens if a copy gets leaked to the press. Those things happen in Washington, you know."

The best Flannery could do was threaten retribution. "You keep those goddam tapes confidential, you sonovabitch, or you'll wish you never made them."

"Keep in touch, Joe. And have a nice day." Majors hung up without waiting for a response.

+++++++++++++++++++++++++++++++++

"Yes, I do," Decker responded to the oath, and then he was instructed to take his seat at the witness table before the panel of Senators.

"Thank you for coming in this afternoon, Mister Decker," Senator Flannery began. "As you may be aware, Rear Admiral Sullivan answered many of the Committee's questions this morning, but there are one or two things that you may be able to clarify for us. By the way, I note that Admiral Sullivan is in the audience, as he advised us this morning he intended to be. Good afternoon, Admiral."

"Good afternoon, Senator. I had hoped to just be a mouse in the corner, but the uniform must have given me away," Sullivan said.

"You are always welcome here, Admiral Sullivan, and you will never succeed in being unnoticed." These light comments by Flannery in his exchange with Sullivan relieved some of the tension in the room. Then back to Decker, "Do you have a prepared statement for the record, Mister Decker?"

"No, Mister Chairman," Decker replied. "I must apologize that I don't. You see, I am not aware of why I was subpoenaed to testify, so I didn't quite know what to prepare for you, but I'll be more than pleased to answer any questions you may have. I do have one matter that I'd like to bring to the Committee's attention, however, with your indulgence."

"Go right ahead, Mister Decker." Flannery spoke with the cordiality that he always displayed in his role as chair of this newsworthy gathering.

"Thank you sir. I'd just like the committee to know that it is not necessary to issue a subpoena to have me appear if you need my assistance in the future, perhaps sparing you a bit of effort. Just a phone call will do." Decker's point was made in a non-argumentative manner, but made nonetheless. He believed it important for the Committee, and the press, to know that he was here voluntarily even though a subpoena had been served.

"We'll keep that in mind, Mister Decker." Flannery agreed, but feeling that his authority had been challenged, elaborated. "We try to use our subpoena power sparingly, and only on those occasions in which there is some uncertainty about a witness's enthusiasm for appearing before us. I trust that we have not offended you."

"Not at all Senator. I'm honored to be here. How can I be of assistance to the committee?" Decker was accustomed to asking questions rather than answering them and could not resist doing so on this occasion, keeping in mind Ray Sycamore's advice to appear cooperative.

Flannery proceeded directly to the issue he wanted to raise, and the occasion of Decker's testimony provided just the right atmosphere in which to raise it. His intended accusation of the SEANET program's duplicity having been squandered in the morning session with Admiral Sullivan, he adopted a new strategy based on what he had learned since that time. His doing so was a great surprise to those who knew about the issue as well as to those who didn't.

"I mentioned at the opening of this morning's session that there was a matter of considerable urgency regarding the SEANET program that needed our attention." This was the moment everyone had waited for, as Flannery was about to disclose his "urgent matter." The room was dead quiet as he proceeded. "It has come to my attention recently that there may have been some industrial misconduct during the acquisition phase of the SEANET program, and it is to my great regret that some of my constituent corporations may have been participants. Do you have any information concerning that matter, Mister Decker?"

Decker was caught completely off guard. He hesitated, but only briefly, as he considered his response. The Crenshaw-Lowry conspiracy had diverted Flannery's attention from SEANET program execution as intended, but it had also, unexpectedly, provided him a new issue. By being the first to raise it, he was making it his own issue for which he would take full credit. Decker thought to himself that Flannery was a superb, if deceitful, politician, and then responded, simply, "I don't believe that I do, Senator."

"Are you being truthful with us, Mister Decker," Flannery accused.

"Of course, I am Senator. Can you be more specific?" Decker answered with his own question, a rhetorical device that he had found useful.

"Before you say anything further, Mister Decker, do you wish to be represented by counsel?" Flannery hardened his accusation.

"Just a minute, Mister Chairman." It was Tennessee's Senator James Tucker who interrupted. Tucker was the opposition party's ranking member in the Senate Armed Services Committee, and he was well aware of Flannery's frequently employed tactic of unjustifiably accusing witnesses of lying. "Mister Decker has responded to a question so broad as to be unanswerable. It is inappropriate for you to accuse him of being untruthful and advising him that he may need to consult legal counsel."

"Well, let's just see how Mister Decker responds," Flannery said.

"My response, Senator," Decker said, "is that I don't understand your question. If you can be more specific, I'll try to answer."

"I have reason to believe, Mister Decker, that two Massachusetts companies, Commware and New England Electronics, exchanged proposal information in violation of a legal agreement with your company, and I also have reason to believe that you knew about it before you came here today. Is that correct, Mister Decker?"

Flannery had phrased the question in such a way that Decker could not avoid a direct answer. "Like you, Senator, I have some suspicions that such an exchange may have taken place, but I am not certain that it did, so I hesitate to make the accusation. If you like, I can provide some background."

"Please tell us what you can, Mister Decker, and remember that you are under oath," Flannery continued.

"Just a damn minute, Mr. Chairman," Tucker interrupted once more. "It was not necessary to remind Mister Decker that he is under oath. When you do that, you are suggesting that the witness has been untruthful. You don't have any reason to suggest that he is, so would you please refrain from using that device?"

Flannery ignored the interruption and readdressed himself to Decker. "Please give us your answer, Mister Decker."

"I'd like to thank Senator Tucker for coming to my defense, Mister Chairman." Decker began. "He made the point, but I would like to make it for myself. It doesn't matter to me whether I am under oath or not. I would give the same answer in either event. I resent the implication that I have been untruthful, and your position on that side of the witness table rather than this side does not give you the right to make an accusation without justification. That said, and I hope, heard, do you want to hear my answer, or do you just want to insult my integrity before I have a chance to give it?

"I will ask the questions here, Mister Decker." Flannery was unsure how to proceed with Decker, but decided to back down somewhat since he appeared to have more backbone than most. "Will you please just give us your answer to my question?"

"Of course," Decker replied, now calmly. "I have worked on SEANET only about six weeks, so I was not a participant in the proposal phase. However, when I began to look into the program's financial status, I noticed something unusual in our subcontractor Commware's cost data. Without boring you with the details, let's just say that they were overrunning their subcontract about the same amount uniformly over all their tasks. That's unusual. Ordinarily there are some ups and some downs that average out to some median level over time. That wasn't happening. I brought these data to the attention of Commware's Program Manager, Mister Harold Laudermilk, who did not seem surprised. I'm told that he, in turn, discussed my concerns with his management, and in a matter of days he was let go by Commware. I met with Mister Laudermilk just two days ago, and at that meeting he provided some information that leads me to the conclusion that Commware had mistakenly provided a proposal to Tellonics that had actually been intended for our competitor, New England Electronics. That proposal was some fifteen percent lower than the proposal intended for Tellonics. If that is what happened, then it would explain the unusually uniform overruns, so I tend to believe it. Commware had an exclusive teaming agreement with Tellonics, so if they provided proposal data to New England Electronics, that would have been in violation of the agreement. However, I can't prove it. It's just a suspicion."

"So you are testifying that Commware and New England Electronics entered into a conspiracy to defraud the Government. Is that your testimony Mister Decker?" Flannery was determined to get a commitment from Decker.

"That is not my testimony, Mister Chairman. I'm certain that my testimony was recorded, so perhaps you would like to hear it read back. If not, then let me paraphrase it for you. I suspect, but I cannot prove, that Commware gave proposal data to New England Electronics in violation of their teaming agreement with Tellonics. I said nothing about a conspiracy or defrauding the Government. Those are your words, Senator, not mine." Decker would not stand for being misquoted or for having his testimony exaggerated.

"What other conclusion could one reach from your testimony, Mister Decker?" Flannery persisted.

"You may reach whatever conclusion you like, Mister Chairman. My testimony contains allegations based on what I have heard from others, but it does not conclude anything." Decker continued to resist Flannery's now transparent attempt to portray his testimony as something more than it was.

"I believe Mister Decker has told us what he knows and all that we need to know, Mister Chairman," Tucker interrupted once more. "If you have further questions for him, please ask them, and let's move on."

"Mister Decker, I do not care for your attitude, and I do not believe your testimony is complete. I think you are holding something back, but as you say, I can't prove it, so we'll just have to look into that outside this hearing room. After that, we may indeed have you back here – with just a phone call, as you recommended." Then addressing himself to the other committee members, "Do the Senators have further questions for Mister Decker?"

"I do have a few questions for Mister Decker, with the chair's permission. It was Tucker again.

"You have the floor, Senator Tucker," Flannery said.

Tucker said, "I hear that there have been some attempts on your life, Mister Decker. Would you enlighten us about that?"

Decker was relieved that Flannery was finished with him and a more friendly voice was being heard. He replied, "I'll tell you what I know, Senator, which is not very much, I'm afraid. The matter is still under investigation by the Orange County Sheriff's Department, assisted by the FBI and the ATF. I have been requested not to disclose certain details that might harm the investigation, but I can tell you what I believe at this time. These did appear to be, as you said, attempts on my life. The first, the bombing of my office, was unsuccessful because I was not present when the bomb exploded. My Administrative Assistant, Erik Hansen, was seriously injured, but he is expected to make a full recovery, I'm happy to report.

"The second attempt occurred as my company's security enhanced car, wisely provided for my protection, entered the parking garage at my townhouse. A gunman in an SUV fired several shots at us, and my bodyguard/driver responded in kind. The gunman was seriously wounded and is still in the hospital – I don't know his condition. His name is Patrick Riley, and I believe that he acted alone. He appears to be a disgruntled employee, a design draftsman, and by putting two and two together, we discovered that he had sabotaged some critical engineering drawings and caused a major production delay. That is being corrected as we speak. That's all I can tell you, and it's most of what I know."

Decker did not disclose the fact that Riley had made several phone calls to Senator Flannery's hot line because Detective Meisner had given him the information in confidence. It would have been a joy to bring it up, to ask the Senator why an assassin would be calling his hotline, but that question would have to wait for another day.

"Thank you for the update, Mister Decker, and I certainly hope that you are right about your attacker having acted alone. Take care of yourself. We need people like you working for us." Following those reassuring words of Senator Tucker, Decker was excused by the Chairman, but with no further comment on his part.

+++++++++++++

Arnold Tell was midway between Massachusetts and California when Susan Anders called the Sovereign's cell phone. "I've got Jack Decker on the line, Arnie. He just finished his testimony and wants to talk to you. I watched it on CSPAN in your office."

"I'd have watched it with you if I'd been there," Tell said. "How'd he do?"

"He did very well. He didn't take any crap from that Senator Flannery, but he was forced to tell them about the Commware deal – the 'alleged' Commware deal he made clear – with New England Electronics. Senator Tucker was great when he came to Jack's defense after Flannery accused him of lying under oath. I'd say he came out the winner and Flannery the loser. I'll put him on." Anders transferred the call to Tell.

"Susan tells me you escaped with your scalp still attached. Tell me all about it." Tell expected that CSPAN would rerun the testimony later that night, but he wanted to hear it first hand from Decker.

Decker gave his accounts of his meeting with Blaisedale and of the hearing. He explained how Flannery had taken the replanning issue off the table as a result of Sullivan's morning testimony and then made the Crenshaw-Lowry deal his own newly discovered issue. When confronted, he couldn't deny it, Decker told Tell.

After answering a few questions from Tell, Decker changed the subject. "How'd your conversation with Crenshaw go, Arnie?"

Tell gave him a brief summary. "The sonovabitch caved on the spot. There won't be any more proxy fight from that bastard, and he's resigning from the board. I told him that he'd

322

better get to that Senator to take the heat off of you if he wanted to keep his deal with New England Electronics quiet. Apparently, he didn't succeed. The whole thing seems to have backfired on him, and rightly so. Now I don't have to worry about keeping that business a secret.

"What about the Air Force General? How'd that meeting go?" Tell asked.

Decker answered, "It went fine, in fact you probably remember him. He was Major Joe Scott on the Northwind Program; now he's a three star general."

"Yeah, I remember him. I'll be damned. Small world, isn't it?" Tell was surprised also.

"The good news is he agreed to reschedule the satellite launches. The bad news is it's going to cost you five hundred K." Decker did not hold back the bad news.

"Well, that's damned generous of you Decker," Tell remarked. "Now how the hell are we gonna pay for that?"

"I may need a temporary small loan, Arnie, but I think SEANET can cover it if we don't have any more disasters. I'll work it out with the bean counters and get back to you." Decker hoped that Tell would not question his decision.

He did not. Tell changed the subject. "How about that Riley guy. What's going on there?"

"Nothing new to report, except I'm now convinced that he did act alone. We found out how he intentionally screwed up some microwave drawings that were causing the critical path production delays. We should have it fixed in a week and new boards out of the factory in six weeks. We can live with that." Decker was glad that he had more good news than bad. "And one other thing, before you ask. I'm pretty confident that we have a fix for the satellite software problem. Commware has a new Program Manager, and she seems pretty sharp."

Tell had news of his own to tell Decker. "That's good. In a month she'll be working for you. We're buying Commware, and I'm putting their whole operation under you. Now that you've fixed the bloody SEANET Program, you need something useful to do in your spare time to earn that small fortune I agreed to pay you in a weak moment."

Decker was shocked, but only for an instant. He had learned that Arnold Tell was never to be underestimated and could be expected to do the unexpected. Decker did not agree, however, that SEANET had been fixed. The major problems – the showstoppers -- had been identified and corrected, but there would be more to come in the year ahead. Still, he welcomed the challenge of bringing Commware into the Tellonics Family. With Ken Martin and Milt Karinski running SEANET under his cognizance, he could find the time. Accepting the challenge, he said simply, "Aye, aye, Sir!"

EPILOGUE

The year that followed the Senate Armed Services Committee's SEANET hearing was one of challenge. There were no additional hearings after Jack Decker gave his testimony. Senator Flannery was reported to have lost his appetite for such a small matter and would spend his time on issues of broader impact.

Absent a Federal indictment, Mason Crenshaw and Ralph Lowry were charged by the Attorney General of the Commonwealth of Massachusetts for entering into an unlawful conspiracy. They were indicted and are awaiting trial in Middlesex County Superior Court.

The acquisition of Commware proceeded as planned, purchased with Arnold Tell's personal Tellonics Shares. Upon approval by its Board of Directors, Tellonics purchased Commware from Tell for an equal number of shares held by the corporate treasury. It was absorbed into the Tellonics corporate family and after six months became the new Tellonics Information Systems Division. It has become a stable and profitable operation under the direction of President and General Manager Jack Decker.

Michael Harrison was promoted to Director of Security for Tellonics, reporting to the Director of Human Relations.

Cynthia Robbins was exonerated in the shooting of Patrick Riley. The authorities found that she had acted in self-defense. She replaced Michael Harrison as chief of security for Arnold Tell.

Confronted with irrefutable evidence, Patrick Riley, on advice of counsel, entered into a plea bargain with the Orange County District Attorney. He pled guilty to one count of aggravated assault causing serious injury, one count of assault with a deadly weapon, and two counts of attempted murder. He was sentenced to forty years in prison with eligibility for parole after twenty-five years.

The major SEANET technical problems were brought under control by the end of December. All three satellites were launched on the new schedule agreed to by the Navy and supported by the Pacific Missile Range. The SEANET contractor tests were completed on schedule, and the system was turned over to the Navy's Operational Test and Evaluation force on the first day of May.

A month later, on June first, Jack Decker and Susan Anders were married in Los Angeles. The ceremony was attended by a small group of friends and relatives. Arnold Tell was Decker's best man.

At the Tellonics annual stockholder's meeting on July fifteenth, the Board of Directors re-elected Arnold Tell as its Chairman and continued his appointment as Chief Executive Officer of the corporation.

OPTEVFOR completed its evaluation of SEANET in sixty days time and declared the system suitable for operational deployment. As a result, the Navy entered into negotiations with Tellonics for Low Rate Initial Production (LRIP) of SEANET hardware to begin equipping the fleet.

On the Monday following the conclusion of contract negotiations, Rear Admiral Russell Sullivan sent the following email message to Arnold Tell and Jack Decker.

"As you know we completed our negotiations for SEANET LRIP on Friday afternoon. Today I signed the documents necessary to put the resulting contract into effect. I want to thank you both for everything you and your company have contributed to SEANET. The country owes you its gratitude, which you will most likely not receive. You may take pride, however, in the results of your effort. The U.S. Navy will be better equipped to go in harm's way in order to meet the unknown challenges of the twenty first century."

The message was signed, Russell A. Sullivan, RADM, USN. It was dated September 10, 2001.

THE END